Campylobacter

Campylobacter
Features, Detection, and Prevention of Foodborne Disease

Editor

Günter Klein

Institute of Food Quality and Food Safety
University of Veterinary Medicine Hannover
Hannover, Germany

ELSEVIER

AMSTERDAM • BOSTON • HEIDELBERG • LONDON
NEW YORK • OXFORD • PARIS • SAN DIEGO
SAN FRANCISCO • SINGAPORE • SYDNEY • TOKYO

Academic Press is an Imprint of Elsevier

Academic Press is an imprint of Elsevier
125 London Wall, London EC2Y 5AS, United Kingdom
525 B Street, Suite 1800, San Diego, CA 92101-4495, United States
50 Hampshire Street, 5th Floor, Cambridge, MA 02139, United States
The Boulevard, Langford Lane, Kidlington, Oxford OX5 1GB, United Kingdom

Library of Congress Cataloging-in-Publication Data
A catalog record for this book is available from the Library of Congress

British Library Cataloguing-in-Publication Data
A catalogue record for this book is available from the British Library

ISBN: 978-0-12-803623-5

For information on all Academic Press publications
visit our website at https://www.elsevier.com/

Working together
to grow libraries in
developing countries

www.elsevier.com • www.bookaid.org

Publisher: Nikki Levy
Acquisition Editor: Patricia Osborn
Editorial Project Manager: Jaclyn Truesdell
Production Project Manager: Caroline Johnson
Designer: Victoria Pearson

Typeset by Thomson Digital

Contents

CHAPTER 4 Isolation, Identification, and Typing of *Campylobacter* Strains from Food Samples **61**

Omar A. Oyarzabal, Catherine D. Carrillo

CHAPTER 5 *Campylobacter* Epidemiology—Sources and Routes of Transmission for Human Infection **85**

Diane G. Newell, Lapo Mughini-Gras,
Ruwani S. Kalupahana, Jaap A. Wagenaar

List of Contributors

Thomas Alter
Institute of Food Hygiene, Free University Berlin, Berlin, Germany

Steffen Backert
Department of Biology, Division of Microbiology, Friedrich Alexander University Erlangen/Nuremberg, Erlangen, Germany

Martijn Bouwknegt
Centre for Infectious Disease Control, National Institute for Public Health and the Environment (RIVM), Bilthoven, The Netherlands

Manja Boehm
Department of Biology, Division of Microbiology, Friedrich Alexander University Erlangen/Nuremberg, Erlangen, Germany

Catherine D. Carrillo
Canadian Food Inspection Agency, Ottawa, ON, Canada

Tadhg Ó. Cróinín
School of Biomolecular and Biomedical Science, University College Dublin, Belfield, Dublin, Ireland

Brecht Devleesschauwer
Department of Public Health and Surveillance, Scientific Institute of Public Health (WIV-ISP), Brussels, Belgium

Arie H. Havelaar
Emerging Pathogens Institute, Department of Animal Sciences and Institute for Sustainable Food Systems, University of Florida, Gainesville, FL, United States

Markus M. Heimesaat
Department of Microbiology and Hygiene, Charité—University of Medicine Berlin, Berlin, Germany

Ruwani S. Kalupahana
Department of Veterinary Public Health and Pharmacology, Faculty of Veterinary Medicine and Animal Science, University of Peradeniya, Peradeniya, Sri Lanka

Günter Klein
Institute of Food Quality and Food Safety, University of Veterinary Medicine Hannover, Foundation, Hannover, Germany

Marie-Josée J. Mangen
Centre for Infectious Disease Control, National Institute for Public Health and the Environment (RIVM), Bilthoven; Department of Public Health, Health Technology Assessment and Medical Humanities, Julius Center for Health Sciences and Primary Care, University Medical Center Utrecht, Utrecht, The Netherlands

Lapo Mughini-Gras
Department of Infectious Diseases and Immunity, Faculty of Veterinary Medicine, Utrecht University, Utrecht; National Institute for Public Health and the Environment (RIVM), Centre for Infectious Disease Control (CIb), Bilthoven, The Netherlands

Diane G. Newell
Department of Infectious Diseases and Immunity, Faculty of Veterinary Medicine, Utrecht University, Utrecht, The Netherlands; School of Veterinary Medicine, Faculty of Health and Medical Sciences, University of Surrey, Guildford, United Kingdom

Sati Samuel Ngulukun
Bacterial Research Department, National Veterinary Research Institute, Vom, Plateau State, Nigeria

Omar A. Oyarzabal
University of Vermont Extension, St. Albans, VT, United States

Felix Reich
Institute of Food Quality and Food Safety, University of Veterinary Medicine Hannover, Foundation, Hannover, Germany

Nicole Tegtmeyer
Department of Biology, Division of Microbiology, Friedrich Alexander University Erlangen/Nuremberg, Erlangen, Germany

Jaap A. Wagenaar
Department of Infectious Diseases and Immunity, Faculty of Veterinary Medicine, Utrecht University, Utrecht; Central Veterinary Institute of Wageningen UR, Lelystad; WHO-Collaborating Center for Campylobacter and OIE-Reference Laboratory for Campylobacteriosis, Utrecht/Lelystad, The Netherlands

Preface

Campylobacter represents a special challenge within the area of foodborne pathogens. From its physiological characteristics, it presents an unlikely pathogen, as it requires conditions during storage that are normally avoided during preservation. For example, a microaerobic milieu, required by *Campylobacter*, is a disadvantage for each foodborne pathogen. On fresh meat, such an atmosphere is rarely found. It furthermore needs specific niches to survive. Freezing is also an effective reduction mechanism for *Campylobacter*, which is normally widely applied, as is drying. Finally, multiplication in food is not possible; in this respect *Campylobacter* is more like viruses than bacteria. But it is still the most important and prevalent bacterial foodborne infection worldwide, so it must have mechanisms not only to survive, but also to infect people via the food chain. It is poultry meat in particular that represents a niche for *Campylobacter* to survive, that is, with high moisture during processing, skin that is less exposed to oxygen, and in many countries marketed as a fresh product (without freezing). In addition, handling mistakes by the consumer (undercooking or crosscontamination) lead to the high exposure and high numbers of infections. On the other hand, mitigation strategies are difficult to implement, and some are not feasible at all, such as vaccination—so far. Especially on the level of primary production, a broad spectrum of methods can be applied, from well-established ones such as biosecurity, to more experimental ones such as bacteriophage therapy. This is also the reason that makes this book different from others in the series "Features, Detection, and Prevention of Foodborne Disease." This is reflected by the chapters ranging from the specifics of human campylobacteriosis (sporadic disease vs. outbreaks and sequelae), to the impact on public health (with DALY, etc.), the basics on taxonomy and physiological characteristics, as well as the classical phenotypic and more advanced molecular detection and typing methods; furthermore, the specific epidemiology with animal and environmental reservoirs, including source attribution, the preharvest to postharvest mitigation and prevention strategies, and finally legal aspects, including microbiological criteria in different countries. This is all very different from *Salmonella*, for example, and requires knowledge from all disciplines in the field. Consequently, we have well-known and very experienced authors from Europe, North America, and Africa, who contributed to the book. This book will be of interest, therefore, to medical doctors, veterinarians, public health officials, company scientists, and interested microbiologists working in the field, as it represents a comprehensive overview of the difficult issue of *Campylobacter* spp. For me, personally, it was a pleasure to work on it, as well as with the colleagues responsible for the different chapters, and with the publishing house.

Günter Klein (Editor)
Hannover, Germany
July 2016

Human campylobacteriosis

1

Steffen Backert*, Nicole Tegtmeyer*, Tadhg Ó Cróinín**, Manja Boehm*, Markus M. Heimesaat†

Department of Biology, Division of Microbiology, Friedrich Alexander University Erlangen/Nuremberg, Erlangen, Germany; **School of Biomolecular and Biomedical Science, University College Dublin, Belfield, Dublin, Ireland; †Department of Microbiology and Hygiene, Charité—University of Medicine Berlin, Berlin, Germany

1.1 INTRODUCTION

The gastrointestinal (GI) tract is one of the largest and most important organs in humans. In a normal adult male, the GI tract is about 6.5m long, and is covered by a layer of intestinal epithelial cells displaying a surface of about 400–500 m². This epithelium exhibits not only crucial absorptive and digestive properties, but also represents an efficient barrier against the commensal microbiota as well as foodborne pathogens (Eckburg et al., 2005). Multiple foodborne and waterborne infections of the human GI tract are responsible for very high rates of morbidity and mortality. According to publications by the World Health Organization (WHO), these infections kill an estimated number of 2.2 million people every year (WHO, 2014). *Campylobacter* has been recognized as the leading cause of bacterial gastroenteritis worldwide (Wassenaar and Blaser, 1999; Young et al., 2007; Nachamkin et al., 2008; Poly and Guerry, 2008). The *Campylobacter* genus consists of a large and diverse group of bacteria, currently comprising more than 30 species and subspecies (http://www.bacterio.net/; see Chapter 3). The most prevalent subspecies with regard to human infection are *Campylobacter jejuni* and *C. coli*. These microbes are typical zoonotic pathogens and members of the ε-proteobacteria. They carry a relatively small genome that is a singular, circular chromosome of 1.59–1.77 Mbp in size, with an average G + C ratio of 30.3–30.6%. The high gene content of 94–94.3% makes it one of the densest bacterial genomes sequenced to date (Parkhill et al., 2000; Fouts et al., 2005; Hofreuter et al., 2006). These Campylobacters are quite fastidious in vitro, growth is limited to microaerobic atmosphere and temperatures of 40–42°C, and requires nutrient-rich media under laboratory conditions (Stingl et al., 2012). Its catabolic capability is highly restricted, which is in contrast to *Salmonella* serovar Typhimurium, and other enteropathogenic bacteria because several genes for common pathways for carbohydrate utilization are either missing or incomplete (Hofreuter, 2014). Despite these metabolic limitations, *C. jejuni* can efficiently colonize various animal hosts as a commensal intestinal inhabitant. During infection of humans, *C. jejuni* enters the host intestine via the oral route and colonizes the GI tract. In this regard, *C. jejuni* appears to be tremendously successful in competing with the intestinal microbiota (Masanta et al., 2013). An infectious dose of a few hundred bacteria is sufficient to overcome the so-called "colonization resistance barrier" in humans, and can lead to campylobacteriosis, but long-term sequelae are also known. The molecular mechanisms underlying disease development by *C. jejuni* infections have begun to emerge in recent years, and will be discussed in this Chapter.

1.2 RISK FACTORS FOR *C. jejuni* COLONIZATION IN ANIMAL HOSTS

Campylobacters were first recognized about 100 years ago as the causative agent of infections in sheep (McFaydean and Stockman 1913). Today, we know that *Campylobacter* spp. are widespread in the natural environment, and can survive for long periods of time outside and inside of a given host (Wassenaar and Blaser, 1999; Young et al., 2007; Nachamkin et al., 2008; Poly and Guerry, 2008). Campylobacters can be frequently found in surface water areas, and are part of the natural intestinal microbiota of a wide range of wild and domestic birds, especially poultry, as well as various agriculturally important mammals. The estimated *Campylobacter* spp. prevalence in poultry and nonpoultry farm animals depends on the season of the year, age of the animals, size and type of flock or herd, husbandry practice, diet, and geographical region (Nachamkin et al., 2008). The flexibility of the *C. jejuni* genome with the presence of highly mutable sites in various genetic loci seems to play a crucial role in its rapid adaptation to a novel host (Sheppard and Maiden, 2015).

A series of risk factors have been reported to be involved in the environmental transmission of Campylobacters to poultry flocks, and these factors appear to be intimately linked to each other. For example, the increased temperature during the summer season promotes the presence of potential *Campylobacter*-carrying vectors such as flies and rodents in poultry farms, while increased rainfall can produce water puddles and ditch water reservoirs in which *C. jejuni* can accumulate, persist, and transmit to new hosts (Jorgensen et al., 2011; Bronowski et al., 2014).

The likelihood of a chicken farm becoming colonized increases over time during rearing, resulting in an average of 60–80% of flocks becoming carriers of *Campylobacter* spp. at the time of slaughter. Studies have shown that most poultry flocks become colonized after 2–4 weeks (Herman et al., 2003; Potturi-Venkata et al., 2007; van Gerwe et al., 2009). The initial colonization of young chickens occurs probably by horizontal transmission from the natural environment, whereas vertical transmission from breeder hens or carryover of infection from a positive flock to a new flock in the same farm, after cleaning and disinfection, are reported to be implausible (Herman et al., 2003; Bronowski et al., 2014). Other studies have shown that various persistent *C. jejuni* isolates in the outside environment can be responsible for repeated infection of rotating poultry flocks (Wedderkopp et al., 2003). Colonized livestock and free-living animals represent important risk factors for the transmission of *C. jejuni* to poultry flocks as the bacterial genotypes from cattle, pigs, and laying hens present at particular farms are similar (Ellis-Iversen et al., 2009; Allen et al., 2011). Taken together, the origin of *Campylobacter* colonization in poultry farms is highly complex, and only a combined scheme of various hygiene practices in each of the discussed sections may help to reduce the burden of *Campylobacter* colonization in the future.

1.3 TRANSMISSION OF CAMPYLOBACTERS TO HUMANS

Transmission of *Campylobacter* spp. to humans occurs most commonly by consumption and handling of various kinds of foods of animal origin whose carcasses were contaminated by the bacteria during slaughter and further processing (Bronowski et al., 2014; Kaakoush et al., 2015). In developed countries, handling, preparation, and consumption of contaminated chicken products is reported to be the main source of human infections. In Germany, the annual incidence of reported human *Campylobacter* infection cases was 63/100,000, with a total of about 52,000 cases in 2006, and increased until 2014 with an annual incidence of 78.9/100,000, and a total of about 71,000 cases. According to the Robert Koch Institute's statistical report, this constituted 44% and 71% of all reported intestinal bacterial infections, respectively (Fig. 1.1). However, the true number of campylobacteriosis cases in Germany and other countries is probably significantly higher (Stingl et al., 2012).

Frequent consumption of chicken meat products reduces the risk for illness after recent chicken consumption. This suggests that partial immunity could be developed,

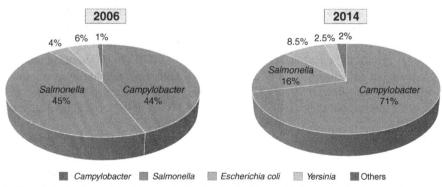

FIGURE 1.1 The Annual Incidence of Intestinal Infections with Bacterial Pathogens in Germany

These data, in accordance with the German Disease Statistics Reports for the years 2006 and 2014, were published by the Robert Koch Institute, Berlin, Germany (http://www.rki.de/EN/Home/homepage_node.html)

allowing protection against *Campylobacter* spp. (Havelaar et al., 2009). Certain *Campylobacter* strains in chicken meat can be linked frequently to human cases of campylobacteriosis, as evidenced by genetic typing methods (see Chapter 5 of this book). However, the overall genotypic diversity between strains demonstrates that there are also other sources contributing to disease in humans. Poultry and nonpoultry farm animals can contribute to campylobacteriosis in humans. For example, the contamination of raw milk and surface water at the farm as well as colonization of poultry flocks in these farms, represent risk factors for transmission of *C. jejuni* to humans (Bronowski et al., 2014; Kaakoush et al., 2015). In addition, direct contact of persons with cattle or pets, in particular puppies with diarrhea, were reported as possible routes of infection (Kwan et al., 2008; Bronowski et al., 2014). Therefore, cattle and their direct environment also represent potential vectors for zoonotic *C. jejuni* isolates.

Moreover, drinking water has been implicated as a possible source for human illness, although, in developed countries, waterborne infection with *Campylobacter* in humans is not very likely. Finally, freshly harvested vegetables can become contaminated with Campylobacters directly at farms, or during transport to, and processing in, food factories. Thus, raw vegetables are another important source of infection, and were reported to represent the second important risk factor after the consumption of contaminated chicken meat (Kwan et al., 2008; Verhoeff-Bakkenes et al., 2011; Bronowski et al., 2014).

1.4 PARAMETERS INFLUENCING HUMAN COLONIZATION AND DISEASE

1.4.1 THE ROLE OF INTESTINAL MICROBIOTA AND COLONIZATION RESISTANCE

The human GI tract is inhabited by myriads of commensal microbes that are collectively referred to as the "intestinal microbiota" (Yoon et al., 2015). The intestinal microbiota is composed of different bacterial phyla and bacteriophages as well as archaea, yeast, and filamentous fungi (Eckburg et al., 2005). The different bacteria in the microbiota are consistent from childhood to adulthood, including the predominant *Escherichia*, *Bacteroides*, *Bifidobacterium*, *Lactobacillus*, *Clostridium*, and *Klebsiella* species, but their number can differ, and is not always constant (Masanta et al., 2013).

The human body benefits greatly from the intestinal microbiota by controlling various processes such as the proper development of the immune system, fermentation of complex carbohydrates to absorbable short chain fatty acids, the detoxification of some harmful substances such as bilirubin and bile acids, as well as establishing a form of defense barrier against intruding pathogens. This phenomenon, also known as colonization resistance, represents a physiological process applied by commensal microbiota that counteracts pathogens from causing intestinal infections

(Yoon et al., 2015). Various parameters, including geographical location, metabolic activity, food consumption, antibiotic exposure, and others, can affect the intestinal colonization resistance. For example, organic acids produced by the metabolism of intestinal microbiota can change the gut pH, which inhibits the growth and spread of pathogenic *Escherichia coli* and *Salmonella* spp. (Cherrington et al., 1991). In addition, the intestinal microbiota uses most oxygen available in the intestine, favoring an anaerobic microenvironment that impairs the capability of various enteric pathogens from colonization (Marteyn et al., 2011).

Remarkably, some metabolic compounds of the microbiota can support *C. jejuni* in colonizing the human gut, and even in entering intestinal epithelial cells. For example, the major short chain fatty acids generated by microbiota are acetate and lactate (Macfarlane and Macfarlane, 2003), and *C. jejuni* was shown to utilize both metabolites as a carbon source (Wright et al., 2009; Thomas et al., 2011). Consequently, acetate and lactate availability in the human gut promotes colonization of *C. jejuni*. In addition, *C. jejuni* was described to utilize bile salts, for example, for the synthesis and secretion of various invasion antigens such as the so-called Cia proteins (Rivera-Amill et al., 2001) that support infection, as discussed later. Furthermore, the bile salt metabolism of intestinal microbiota also produces fumarate. *C. jejuni* has been shown to encode the methyl-menaquinol fumarate reductase enzyme that permits the bacteria to use fumarate as an electron acceptor under oxygen-limiting conditions (Weerakoon et al., 2009; Guccione et al., 2010).

As mentioned previously, several factors are known to control intestinal colonization resistance. Recent studies have demonstrated that a diet-triggered shift of the intestinal microbiota enhanced the susceptibility for *C. jejuni* colonization in gnotobiotic mice (Bereswill et al., 2011). When the animals were fed with a human western-style diet for 6 weeks, the microbiota composition changed to more human-like microbiota, as compared to conventional control mice. Remarkably, these "humanized" mice were strongly susceptible to *C. jejuni* infection, whereas corresponding control mice exhibited a murine microbiota composition, and were infection resistant (Bereswill et al., 2011). Interestingly, the western-style diet-fed mice exhibited higher numbers of *Clostridium/Eubacterium* spp. and *E. coli*, and lower *Enterococcus* and *Lactobacillus* spp. colonization, compared to conventional murine diet-fed animals. These experiments support the view that individuals consuming a western diet may be more susceptible to *C. jejuni* infections, as compared to those on a rather low-fat plant, polysaccharide-rich diet (Masanta et al., 2013).

The role of commensal *E. coli* in abrogating intestinal colonization resistance against *C. jejuni* has been shown previously (Haag et al., 2012). When intestinal *E. coli* loads of adult mice harboring a conventional microbiota were artificially elevated by serving a commensal murine *E. coli* strain in drinking water, mice could be stably infected with *C. jejuni* at high rates, while the control animals could not. Future studies will dissect the importance of distinct bacterial species, and the crosstalk of commensal microbiota, *C. jejuni*, and innate immune responses by the host.

1.4.2 HOST EPITHELIAL CELL ADHESION BY *C. jejuni*

After entering the human GI tract, *C. jejuni* first interact with the mucus layer before binding to the epithelial cells of the intestine, and this attachment, as well as cellular invasion and transmigration, appear as necessary requirements for successive colonization and pathogenesis by these bacteria (Dasti et al., 2010; Alemka et al., 2012; Backert et al., 2013). A model for the individual steps of infection is shown in Fig. 1.2. Adherence of *C. jejuni* to target cells involves various adhesion factors (named adhesins), and their corresponding host cell receptors. The binding of these adhesins is suggested to be fundamental for efficient interaction of the bacteria with host target cells, and may be directly participating in the invasion process. Numerous adhesins have been described, including the *Campylobacter* adhesin to fibronectin (CadF), fibronectin like protein A (FlpA), periplasmic binding protein 1 (PEB1), jejuni lipoprotein A (JlpA), *Campylobacter* autotransporter protein A (CapA), major outer membrane protein (MOMP), and p95 (reviewed by Ó Cróinín and Backert, 2012). Probably the best studied binding factor is CadF, a 37-kDa outer membrane protein that permits the attachment of *C. jejuni* to the extracellular matrix

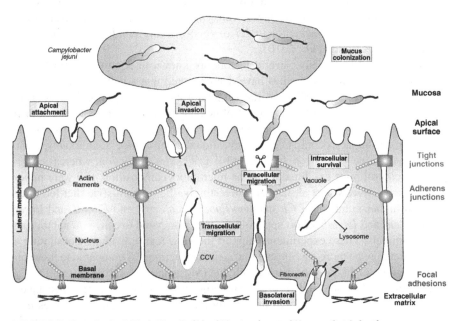

FIGURE 1.2 Hypothetical Model for *C. jejuni* Mechanisms of Human Gut Infection

The pathogen can interact with, invade into, transmigrate across, and survive within polarized intestinal epithelial cells, as indicated.

Adapted from Backert, S., Hofreuter, D. 2013. Molecular methods to investigate adhesion, transmigration, invasion and intracellular survival of the foodborne pathogen Campylobacter jejuni. *J. Microbiol. Methods 95, 8-23. For more details, see the text*

protein fibronectin (Konkel et al., 1997, 1999, 2005). An isogenic $\Delta cadF$ mutant exhibited significantly reduced levels of adherence to, and invasion into, INT-407 intestinal epithelial cells (Monteville and Konkel, 2002; Monteville et al., 2003; Krause-Gruszczynska et al., 2007), and was incapable to colonize chickens (Ziprin et al., 1999). CadF seems to be essential for the uptake of *C. jejuni* at the basolateral host cell surface (Monteville and Konkel, 2002). Numerous reports studying the infection of INT-407 cells showed that *C. jejuni* can bind to fibronectin by CadF (Krause-Gruszczynska et al., 2011; Boehm et al., 2011; Eucker and Konkel, 2012). FlpA represents a 46-kDa factor that also contributes to the attachment of *C. jejuni* to fibronectin, and consequently invasion of epithelial cells. The interaction of FlpA with fibronectin is a dose-dependent process, and a $\Delta flpA$ mutant bound to INT-407 cells significantly less, when compared to the corresponding wild-type bacteria (Konkel et al., 2010; Eucker and Konkel, 2012). These studies indicate that CadF and FlpA probably act cooperatively in mediating *C. jejuni* attachment to the host cells via fibronectin and its receptor integrin. In addition, JlpA is a 43-kDa protein and mutation in the *jlpA* gene, resulted in an 18–19.4% reduced adherence of *C. jejuni* to cultured HEp-2 cells (Jin et al., 2001, 2003). Moreover, it was shown that a pretreatment of HEp-2 cells with purified JlpA led to a decrease of attachment by *C. jejuni* in a dose-dependent manner (Jin et al., 2001). In addition, PEB1 is a 28-kDa protein, and inactivation of the *peb1A* gene decreased adherence of *C. jejuni* to cultured HeLa cells, and prevented colonization of mice (Pei et al., 1998). MOMP (Moser et al., 1997), p95 (Kelle et al., 1998), and CapA (Ashgar et al., 2007) were also reported to operate as bacterial adhesins, but they were not yet examined in much detail. However, it should be noted that some of these proposed *C. jejuni* adhesins are controversial in the literature, a fact that requires clarification in the future (reviewed by Ó Cróinín and Backert, 2012).

1.4.3 CELLULAR INVASION AND TRANSMIGRATION OF *C. jejuni*

A major disease-associated feature of *C. jejuni* is its capability to enter host tissues. Early electron microscopic investigations of intestinal biopsy samples from infected patients, and studies of cultured nonphagocytic cell lines in vitro revealed that *C. jejuni* can invade human intestinal epithelial cells (van Spreeuwel et al., 1985; Oelschlaeger et al., 1993). In general, epithelial cell invasion by *C. jejuni* proceeds in a very specific fashion, and has been described as a primary mechanism leading to tissue damage and pathogenesis (Fig. 1.3). Using specific inhibitors, multiple studies have shown that *C. jejuni* can enter cultured cells by microtubule-dependent (actin-filament-independent) and/or actin-filament-dependent (microtubule-independent) mechanisms (reviewed by Ó Cróinín and Backert, 2012). Various lines of evidence indicated that entry of *C. jejuni* into INT-407 and other cell lines is accompanied by the activation of small Rho GTPases that are required for the invasion process (Krause-Gruszczynska et al., 2007; Eucker and Konkel, 2012). In particular, the GTPase members Cdc42 and Rac1 (but not RhoA) are activated upon infection, and play a significant role in bacterial uptake. The signaling cascades resulting in the Cdc42 and Rac1 activation

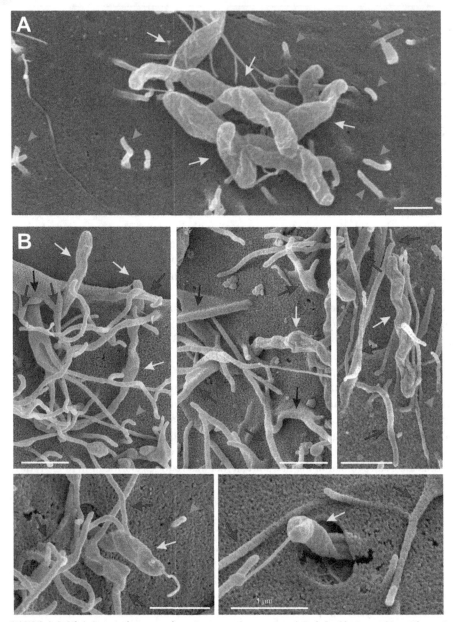

FIGURE 1.3 High Resolution scanning electron microscopy of *C. jejuni*-induced filopodia and membrane ruffling formation, followed by bacterial invasion

(A) Infection of integrin receptor knockout control cells with wild-type *C. jejuni* strain 81–176 [*white arrows* (*yellow arrows* in the web version)] for 6 h revealed bacterial attachment to the cell surface with short microspikes [*white arrowheads* (*green arrowheads* in the web version)] present, but membrane dynamics events or invasion were rarely seen. Similar observations were made with infected fibronectin$^{-/-}$ or FAK$^{-/-}$ cells.

(B) Infecting *C. jejuni* in wild-type cells were regularly associated with long filopodia [*grey arrows* with black lining (*blue arrows* in web version)], membrane ruffling (*black arrows*), as well as elongated microspikes [*white arrowheads* (*green arrowheads* in web version).

Krause-Gruszczynska, M., Boehm, M., Rohde, M., Tegtmeyer, N., Takahashi, S., Buday, L., Oyarzabal, O.A., Backert, S., 2011. The signaling pathway of C. jejuni induced Cdc42 activation: role of fibronectin integrin beta1, tyrosine kinases and guanine exchange factor Vav2. Cell Commun. Signal. 9, 32 encompass CadF binding to fibronectin, and involve various receptors (PDGF and EGF receptors, integrins), cytosolic kinases (PI3-kinase, Src and FAK), and guanine exchange factors (Tiam-1, Vav2, and DOCK180). Experiments applying knockout cell lines, electron microscopy, and other methods have demonstrated that *C. jejuni* stimulates the CadF→fibronectin→β1-integrin→FAK/Src→PDGFR/EGFR→PI3-kinase→Vav2, and CadF→fibronectin→β1-integrin→FAK→Tiam-1/DOCK180 signal transduction pathways to activate Cdc42 and Rac1, respectively, triggering cytoskeletal rearrangements and, subsequently, invasion (Krause-Gruszczynska et al., 2011; Boehm et al., 2011).

Interestingly, *C. jejuni* does not encode typical disease-related type-III or type-IV secretion systems (T3SS or T4SS), as reported for the highly invasive *Salmonella, Shigella*, or *Bartonella* spp. (Parkhill et al., 2000; Fouts et al., 2005; Hofreuter et al., 2006; Siamer and Dehio, 2015). Instead, the *C. jejuni* flagellum and motility are required for increased host cell entry. Mucosal viscosity plays a crucial role for enhanced *C. jejuni* motility, cell attachment, and invasion (Szymanski et al., 1995). It was demonstrated later that the *C. jejuni* flagellum can also function as a T3SS for the export of proteins that can control bacteria–host interactions (Young et al., 1999; Konkel et al., 1999; Christensen et al., 2009; Barrero-Tobon and Hendrixson, 2012). These secreted factors comprise the so-called flagellar coexpressed determinants (FedA-D) and *Campylobacter* invasion antigens (CiaA-H) (Konkel et al., 1999; Eucker and Konkel, 2012; Barrero-Tobon and Hendrixson, 2012, 2014). However, the *C. jejuni* delivery system and function of the secreted proteins are still weakly investigated. One of the best-characterized Cia proteins is the ~70-kDa protein CiaB. CiaB appears to be necessary for the secretion process itself, and is required for maximal invasion of *C. jejuni* into host target cells (Konkel et al., 1999). Gentamicin protection assays that measure the numbers of intracellular bacteria by quantifying the colony forming units showed significant lower invasion rates of a Δ*ciaB* deletion mutant in human cells, as compared to wild-type *C. jejuni*, and the Δ*ciaB* mutant also revealed reduced colonization in chickens (Ziprin et al., 2001).

Another identified *C. jejuni* protein, FlaC, has been also described to be secreted from the flagellar export apparatus. Mutants of *flaC* are still motile, and express a functional flagellum, but they are defective to enter epithelial cells (Song et al., 2004). These observations suggest that the flagellar export machinery represents a crucial secretory device that could explain its requirement for *C. jejuni* host cell entry. However, another report brought up substantial doubt on the above CiaB importance in the secretion of virulence factors and host cell entry (Novik et al., 2010). Therefore, it is still not understood if CiaB and the flagellar device are involved directly in exporting Cia proteins triggering bacterial internalization or if the observed invasion defects of flagellar deletion mutants are due to lack of flagella-based mobility, and consequently reduced bacterial contact with the host cell.

Recent studies have demonstrated that high temperature requirement A (HtrA), a serine protease, is a novel virulence factor of *C. jejuni* with important roles in adhesion, invasion, and transmigration (Brøndsted et al., 2005; Baek et al., 2011; Boehm et al., 2012; Backert et al., 2013). The bacteria actively secrete HtrA proteins in the extracellular space, where they can hijack host cell proteins (Hoy et al., 2010, 2012; Boehm et al., 2015). Infection experiments in vitro indicated that HtrA can open cell-to-cell junctions between epithelial cells by proteolytic cleavage of the adherens junction protein E-cadherin, followed by paracellular transmigration of the bacteria. More recent studies have shown that purified outer membrane vesicles of *C. jejuni* were able to cleave both E-cadherin and the tight junction protein occludin, and resulted in enhanced levels of bacterial adhesion and invasion in a time- and dose-dependent manner (Elmi et al., 2015). Taken together, invasion and transmigration by *C. jejuni* are highly complex processes that require more studies in future.

1.5 SPORADIC DISEASE AND OUTBREAKS

As a consequence of *C. jejuni* infection in humans, the bacteria colonize the ileum and colon, where they can affect regular absorptive and secretory functions of the GI tract. This eventually leads to intestinal campylobacteriosis disease associated typically with fever, malaise, abdominal pain, and watery diarrhea that frequently includes blood (Wassenaar and Blaser, 1999; Poly and Guerry, 2008). Endoscopic findings are often nonspecific, and reveal friable colonic mucosa with associated erythema and hemorrhage. Histological examination exhibits characteristics of acute self-limited colitis, including neutrophilic infiltrates within the lamina propria.

Symptoms commonly occur within 1–5 days of *C. jejuni* exposure, and can continue for up to 10 days. Most of these infections, however, are self-limited, particularly in healthy persons, although relapse is common. Interestingly, epidemiological studies over the past 10 years have shown a steady increase in the overall prevalence of campylobacteriosis in developed and developing countries (Kaakoush et al., 2015). In developed countries, a steady increase in the number of specific outbreaks of *Campylobacter* infection has also been observed (Crim et al., 2015; Tam et al., 2012). Other studies have suggested that the burden of *Campylobacter* diarrhea in developing countries has been grossly underestimated, with *Campylobacter* the most frequent pathogen detected in a variety of studies focusing on bacterial gastroenteritis in children (Platts-Mills and Kosek, 2014). These increasingly higher levels of incidences of *Campylobacter* infections are thought to be only a fraction of the actual number, due to low levels of reporting of this infection.

Although as stated previously, most infections with *Campylobacter* spp. are thought to be due to the consumption of contaminated poultry products or water (Kaakoush et al., 2015), these are predominantly sporadic infections, and are rarely traced back to the specific source of infection. Outbreaks have been less prevalent in the past, but studies from the developed world have revealed an increase in outbreaks of *Campylobacter* gastroenteritis, in contrast to a gradual decrease in infectious outbreaks by other pathogens such as *Salmonella* spp. (Gormley et al., 2011). Poultry meat and resulting products such as liver paté have been identified as important sources of *Campylobacter* outbreaks, in some countries (Scott et al., 2015; Farmer et al., 2012). In other areas such as Scandinavia, waterborne outbreaks are very common, with *Campylobacter* representing 29% of outbreaks with known etiology over a 15-year period (Guzman-Herrador et al., 2015). Furthermore, some new reservoirs have emerged for *Campylobacter* outbreaks, including unpasteurized or poorly pasteurized milk. The increase in the consumption of raw, unpasteurized milk has led to an increase in milk-associated outbreaks of *Campylobacter* infection in recent years, particularly in developed countries (Hauri et al., 2013; Gould et al., 2013).

1.6 POSTINFECTIOUS SEQUELAE OF CAMPYLOBACTERIOSIS

Following *Campylobacter*-induced disease, postinfectious sequelae might arise in some patients after recovery, and can affect the nervous system (Guillain–Barré syndrome, GBS; Miller–Fisher syndrome, MFS; Bickerstaff encephalitis) and the joints (reactive arthritis, RA). In addition, it is hypothesized that *Campylobacter* spp. infections are associated with the development of inflammatory bowel diseases (IBD) such as Crohn's disease and ulcerative colitis, and irritable bowel syndrome (IBS) (Keithlin et al., 2014). A recent metaanalysis revealed that in 0.07, 2.86, 0.40, and 4.01% of *C. jejuni* infection cases, postinfectious sequelae such as GBS, RA, IBD, and IBS, respectively, developed (Keithlin et al., 2014). Furthermore, in 34–49% of GBS, 44–62% of RA, and 23–40% of IBD cases were seropositive for *Campylobacter*, whereas anti-*Campylobacter* antibodies could be detected in 16–26 % of healthy individuals (Schmidt-Ott et al., 2006; Zautner et al., 2014). The fact that seropositivity for *Campylobacter* was markedly higher than for other suspected pathogens provided further evidence of an important role of *C. jejuni* in the immunopathogenesis of the respective postinfectious diseases.

1.6.1 GUILLAIN–BARRÉ SYNDROME

In 1916, the French neurologists Guillain, Barré, and Strohl described acute areflexic paralysis, and subsequent recovery, in two soldiers (Guillain et al., 1916). Later on, GBS was considered a rather heterogenous disease of the peripheral nervous system. Patients present with acute or subacute symmetrical ascending weakness of the limbs, hypo- or areflexia, and mild to moderate sensory symptoms, such as paresthesia or numbness in a glove- and stocking type manner (Hughes and Cornblath, 2005; Wakerley and Yuki, 2015). The clinical course, severity, and outcomes of disease can vary considerably. The most common GBS subtypes include demyelinating forms such as acute inflammatory demyelinating polyneuropathy (AIDP), as well as axonal forms, for example acute motor axonal neuropathy (AMAN), and acute motor and sensory axonal neuropathy (AMSAN), a severe variant of AMAN (van den Berg et al., 2014). A less common GBS subtype is the Miller–Fisher syndrome (MFS), a triad of ophthalmoplegia, ataxia, and areflexia (Fisher, 1956). Some MFS patients with ophthalmoplegia and ataxia might also present with hypersomnolence, a condition termed Bickerstaff brain-stem encephalitis (Bickerstaff, 1957).

Whereas the majority of GBS patients recover, autonomic (mostly cardiovascular) dysfunction with variable severity can be observed in up to two-thirds of patients, of which approximately 25% of the affected individuals develop respiratory insufficiency requiring mechanical ventilation (van den Berg et al., 2014), and up to 10% of patients die (Alshekhlee et al., 2008).

In about two thirds of GBS cases worldwide, antecedent GI or respiratory infections by *C. jejuni*, Cytomegalovirus, Epstein–Barr virus, *Haemophilus influenzae*, or *Mycoplasma pneumoniae* have been documented (Jacobs et al., 1998). Among these, at least one-third were caused by *C. jejuni* (Islam et al., 2010; Jacobs et al., 1998; Yuki and Hartung, 2012). Even though a strong association between acute infections and GBS exists, the overall risk for developing postinfectious disease is rather low. For example, only 1 in 1000–5000 patients develop GBS within 2 months following *C. jejuni* infection (Nachamkin et al., 1998; Tam et al., 2006), indicating that GBS is mostly a sporadic complication, but occasional outbreaks have also been reported (Jackson et al., 2014).

Molecular mimicry and cross-reactive humoral immune responses in susceptible hosts are critical for GBS pathogenesis. The structural similarity between sialylated lipooligosaccharide (LOS) on the outer membrane of *C. jejuni*, with distinct ganglioside structures of human peripheral nerves such as GM1 and GD1, is a typical example of molecular mimicry, and triggers cross-reactive humoral immune responses leading to nerve damage (van den Berg et al., 2014; Yuki et al., 2004). In AMAN, anti-GM1 and anti-GD1a antibodies bind to respective antigens of the axolemma, located at or close to the node of Ranvier. Subsequent complement activation, formation of membrane attack complexes, and invasion of macrophages lead to myelin degeneration and nerve conduction failure (van den Berg et al., 2014). In AIDP, antibodies against various complexes of glycolipids located on the myelin sheath of the Schwann cells have been described, but their exact role during pathogenesis is still unclear (Rinaldi et al., 2013). Furthermore, with GT1a cross-reacting anti-GQ1b antibodies are associated strongly with MFS and Bickerstaff brain-stem encephalitis (Yuki and Hartung, 2012).

The fact that only 1 in 1000–5000 patients develops GBS within 2 months following *C. jejuni* infection (Nachamkin et al., 1998; Tam et al., 2006) supports the hypothesis that distinct bacterial and/or host related factors may determine the susceptibility for, and outcome of, postinfectious disease in the individual host. Interestingly, cross-reactive antibodies were not produced in uncomplicated *C. jejuni* enteritis (Ang et al., 2002; Kuijf et al., 2010), even though in up to 63% of cases sialylated *C. jejuni* strains were found (Godschalk et al., 2004). A very recent study revealed that, besides the sialylated LOS, distinct capsular types of *C. jejuni* strains including HS1/44c, HS2, HS4c, HS19, HS23/36c, and HS41 are involved in GBS pathogenesis. Furthermore, geographical differences in *C. jejuni* capsular type distribution in GBS patients could be demonstrated (Heikema et al., 2015). Polymorphisms of the *C. jejuni* sialyltransferase gene *cst-II* gene have been shown to be associated with different courses of postinfectious disease. Whereas the Thr51 variant of the *cst-II* gene determines the development of GBS, the Asn51 polymorphism is associated with MFS (Yuki and Koga, 2006). The incidence rates of GBS-related *cst-II* and *neuA* genes (the latter encoding for the enzyme acetyl-neuraminate cytidylyltransferase that is responsible for activation of sialic acid) have been shown to be comparably present in *C. jejuni* strains isolated from different sources, including human, cattle, turkey, or chicken, indicating that a selection for or against GBS associated *C. jejuni* isolates in the environment is rather unlikely (Amon et al., 2012).

Despite the high numbers of isolates from different environmental sources bearing the risk for postinfectious GBS, host factors including genetics and innate as well as humoral immune responses determine critically the individual's predisposition for disease development. Pathogen-host interaction is a key event during the immunopathological cascade leading to GBS. Following binding of *C. jejuni* LOS to siglec-7 (sialic acid binding immunoglobulin-like lectin 7), antigen presenting cells such as dendritic cells (DCs) are activated in a Toll-like receptor (TLR)-4 and CD14-dependent fashion, and secrete proinflammatory cytokines, including TNF-α and type-1 interferons leading to B-lymphocyte proliferation (Heikema et al., 2013; Huizinga et al., 2013; Kuijf et al., 2010). A strong individual TLR-4 dependent DC response to *C. jejuni* LOS has been shown to be a critical host condition for GBS development following *C. jejuni* infection (Huizinga et al., 2015). Genetic polymorphisms might, therefore, impact the innate and adaptive immune responses to *C. jejuni* infection. Until now, valid data regarding the impact of genetic factors on GBS susceptibility are scarce, due to small sample sizes. In association studies, distinct polymorphisms in genes encoding TNF-α or mannose-binding protein C (i.e., *MBL2* gene) were associated with severity and outcome of GBS (Wu et al., 2012; Geleijns et al., 2006). However, genome-wide association studies in larger cohorts are needed to further unravel genetic host factors triggering GBS pathogenesis.

During the past 30 years, intravenous (IV) immunoglobulins and plasma exchange have been propagated as treatment options for GBS, whereas GBS patients do not respond to corticosteroids (Wakerley and Yuki, 2015). Intravenous immunoglobulins are thought to exert beneficial effects by antagonizing pathogenic autoantibodies, and subsequent complement activation, whereas plasma exchange measures aim at removing autoantibodies, proinflammatory immune mediators, and complement from the circulation (Arcila-Londono and Lewis, 2012). Patient-related and socioeconomic factors determine the respective choice of treatment, given that plasma exchange is rather invasive, and requires special equipment. The combination of IV immunoglobulins and plasma exchange does not further improve GBS outcome than application of either measure alone (Hughes et al., 2007; van den Berg et al., 2014). Novel therapeutic approaches target the complement system with monoclonal antibodies and synthetic serine protease inhibitors, but further studies are needed to unravel efficacy and safety (Halstead et al., 2008; Wakerley and Yuki, 2015). However, the management of symptoms and complications of GBS is critical and should be accomplished in a multidisciplinary setting.

1.6.2 **REACTIVE ARTHRITIS AND REITER'S SYNDROME**

Reactive arthritis (RA) is defined as an "arthritis which developed soon after or during an infection elsewhere in the body, but in which the microorganisms cannot be recovered from the joint" (Ahvonen et al., 1969). This postinfectious spondylarthropathy is characterized by inflammatory changes of the joints following GI or genitourinary infections, and is associated strongly with HLA-B27 (Ajene et al., 2013). Together with conjunctivitis and urethritis, RA is a prominent symptom within the clinical triad of Reiter's syndrome (Pope et al., 2007). RA usually develops within 4 weeks postinfection, can be self-limiting, and resolves within 6 months. Up to 63% of affected patients, however, are estimated to develop a specific chronic form of RA (Carter, 2006). Symptoms vary from mild mono- or oligoarthralgia to severe disabling polyarthritis. Predilection sites are the joints of the lower extremities, including knees and ankles, but smaller joints can also be affected (Pope et al., 2007).

Campylobacter, *Salmonella*, and *Shigella* spp. are the most common pathogens associated with RA (Wu and Schwartz, 2008). In 1–5% of *Campylobacter*-infected patients, RA may occur, whereas the annual incidence of this postinfectious complication is estimated to be 4.3 cases per 100,000 individuals (Pope et al., 2007). In a more recent systematic review, a weighted mean incidence of 9 RA cases per 1000 *Campylobacter* infections was calculated (Ajene et al., 2013). Notably, neither defined criteria for RA diagnosis exist to date, nor are definite time points from infection to the onset of RA established (Townes, 2010). Hence, reported cases include a rather broad variety of symptoms that might resemble other spondylarthropathic disorders.

The pathophysiology of RA is largely unknown. It has been hypothesized, however, that the interaction of distinct bacterial species with host HLA-B27 plays a pivotal role in RA development (Pope et al., 2007). HLA-B27-positive individuals have been shown to develop more severe and more frequently chronic disease, as compared to HLA-B27 negative individuals in some studies (Calin and Fries, 1976; Leirisalo et al., 1982). In contrast, no such association could be demonstrated in a population-based survey (Hannu et al., 2002). In genetically predisposed patients, bacterial antigens are thought to translocate from the intestinal tract to extraintestinal compartments, including the joints, via the blood stream. As described for GBS, molecular mimicry and hence cross-reactivity of bacterial antigens with molecular structures of the synovia results in proinflammatory immune responses and, subsequently, to precipitating synovitis (Hill Gaston and Lillicrap, 2003; Pope et al., 2007). Furthermore, genetic host factors including polymorphisms leading to an imbalance of pro- and antiinflammatory cytokine expression contribute to initiation and persistence of disease (Appel et al., 2004; Braun et al., 1999; Butrimiene et al., 2004; Kaluza et al., 2001).

1.6.3 **INFLAMMATORY BOWEL DISEASES**

Inflammatory bowels diseases (IBD) are characterized by chronic intestinal inflammation, with acute episodes and rising incidence worldwide (Molodecky et al., 2012). Crohn's disease and ulcerative colitis can be differentiated according to clinical presentation, abundance of inflammation within the GI tract, endoscopic appearances, and histopathological changes (Sartor, 1995). The interplay among the intestinal microbiota, host immunity, genetic, environmental, and so far undefined factors are involved in the multifactorial etiology of IBD (Ananthakrishnan, 2015). The impact of pathogens including *Campylobacter* spp. in IBD pathogenesis is of current debate. Older studies failed to reveal an association between *C. jejuni* and IBD (Blaser et al., 1984; Boyanova et al., 2004; Weber et al., 1992). More recently, however, *C. jejuni* infection was shown to be associated with an increased risk for development and relapses of IBD (Gradel et al., 2009), whereas other emerging *Campylobacter* spp. such as *C. concisus* have come into focus previously (Kaakoush et al., 2015; Mahendran et al., 2011; Zhang, 2015).

A very recent comprehensive metaanalysis of more than 80,000 individuals (Castano-Rodriguez et al., 2015) revealed that infection with *C. concisus* or *C. showae* increased the risk of IBD development. Both bacterial species commonly inhabit the human oral cavity as commensals (Zhang et al., 2014), and may act as opportunistic pathogens in individuals with a compromised immune system (Engberg et al., 2000). Furthermore, in some healthy individuals and, more frequently in patients with active IBD, multiple different *C. concisus* strains could be detected (Mahendran et al., 2013). Viable *C. concisus* among other *Campylobacter* spp. are hypothesized to translocate from the oral cavity to the lower intestinal tract where they might initiate the cascade of mucosal inflammation (Zhang, 2015). The finding that the prevalence of *C. concisus* is similar in newly diagnosed and relapsed IBD cases, but lower in highly inflamed areas, as compared to adjacent less or even uninflamed tissue sites, further points towards a primary event of *C. concisus* infection in IBD development (Mahendran et al., 2011; Zhang, 2015; Zhang et al., 2009).

Interestingly, a dual role of *Helicobacter* and *Campylobacter* spp. has been proposed in IBD development recently. Whereas, as already stated, the prevalence of *C. concisus* and *C. showae* was higher in IBD patients, a negative association between *Helicobacter pylori* infection and IBD could be determined (Castano-Rodriguez et al., 2015). Given the cross-reactivity between both pathogenic species, anti-*H. pylori* antibodies might confer protective adaptive immunity against subsequent *C. jejuni* infection (Newell et al., 1984; Tindberg et al., 2001) that, in turn, decreases the chance of chronic intestinal inflammation (Castano-Rodriguez et al., 2015). However, the distinct mechanisms and potential roles of distinct *Campylobacter* spp. and/ or pronounced host immune responses in the initiation and perpetuation of IBD need to be further unraveled.

1.6.4 IRRITABLE BOWEL SYNDROME

Up to 17% of patients suffering from IBS believe their symptoms began with an infectious disease (Spiller and Garsed, 2009). Acute infectious gastroenteritis is considered a commonly identifiable risk factor for the development of IBS (Spiller and Garsed, 2009). Other risk factors include smoking, preceding mucosal inflammation, female gender, and depression (Spiller and Garsed, 2009). Studies of

C. jejuni infection outbreaks reported between 7.9 and 13.0% of postinfectious IBS cases (Dunlop et al., 2003; Marshall et al., 2007; Thornley et al., 2001). A recent review and metaanalysis revealed that, in prospective studies, 8.64% of *Campylobacter* infected patients developed IBS, as compared to 0.15% in retrospective studies (Keithlin et al., 2014). The mechanisms underlying postinfectious IBS, however, are unknown so far, but might include host genetic alterations, residual inflammation, persistent enterochromaffin cell hyperplasia, serotonergic and mast cell activation, enteric nerve dysfunction, increased influx of proinflammatory immune cells such as T lymphocytes, alterations in barrier function, and changes in the intestinal microbiota composition (i.e. dysbiosis), as well as in the intraluminal intestinal milieu (Grover, 2014; Spiller and Garsed, 2009).

1.7 FUTURE DIRECTIONS

A detailed understanding of the molecular mechanisms underlying host-pathogen interaction in respective acute human campylobacteriosis and postinfectious sequelae are utmost critical for future translational approaches. For continued advances in the field, experts in microbiology, immunology, molecular biology, genetics, infectious diseases, neurosciences, gastroenterology, and rheumatology need to enhance their collaborative efforts.

REFERENCES

Ahvonen, P., Sievers, K., Aho, K., 1969. Arthritis associated with Y*ersinia enterocolitica* infection. Acta Rheumatol. Scand. 15, 232–253.

Ajene, A.N., Fischer Walker, C.L., Black, R.E., 2013. Enteric pathogens and reactive arthritis: a systematic review of *Campylobacter*, *Salmonella* and *Shigella*-associated reactive arthritis. J. Health Popul. Nutr. 31, 299–307.

Alemka, A., Corcionivoschi, N., Bourke, B., 2012. Defense and adaptation: the complex interrelationship between *Campylobacter jejuni* and mucus. Front. Cell Infect. Microbiol. 2, 15.

Allen, V.M., Ridley, A.M., Harris, J.A., Newell, D.G., Powell, L., 2011. Influence of production system on the rate of onset of *Campylobacter* colonization in chicken flocks reared extensively in the UK. Br. Poult. Sci. 52, 30–39.

Alshekhlee, A., Hussain, Z., Sultan, B., Katirji, B., 2008. Guillain-Barre syndrome: incidence and mortality rates in US hospitals. Neurology 70, 1608–1613.

Amon, P., Klein, D., Springer, D., Jelovcan, S., Sofka, D., Hilbert, F., 2012. Analysis of *Campylobacter jejuni* isolates of various sources for loci associated with Guillain-Barre Syndrome. Eur. J. Microbiol. Immunol. (Bp) 2, 20–23.

Ananthakrishnan, A.N., 2015. Epidemiology and risk factors for IBD. Nat. Rev. Gastroenterol. Hepatol. 12, 205–217.

Ang, C.W., Laman, J.D., Willison, H.J., Wagner, E.R., Endtz, H.P., De Klerk, M.A., Tio-Gillen, A.P., Van den Braak, N., Jacobs, B.C., Van Doorn, P.A., 2002. Structure of *C.jejuni* lipopolysaccharides determines antiganglioside specificity and features of GBS and Miller Fisher patients. Infect. Immun. 70, 1202–1208.

Appel, H., Neure, L., Kuhne, M., Braun, J., Rudwaleit, M., Sieper, J., 2004. An elevated level of IL-10- and TGFβ-secreting T cells, B cells and macrophages in the synovial membrane of patients with reactive arthritis compared to rheumatoid arthritis. Clin. Rheumatol. 23, 435–440.

Arcila-Londono, X., Lewis, R.A., 2012. Guillain-Barré syndrome. Semin. Neurol. 32, 179–186.

Ashgar, S.S., Oldfield, N.J., Woolridge, K.G., Jones, M.A., Irving, G.J., Turner, D.P., Ala'Aldeen, D.A., 2007. CapA, an autotransporter protein of *Campylobacter jejuni* mediates association with human epithelial cells and colonization of the chicken gut. J. Bacteriol. 189, 1856–1865.

Backert, S., Boehm, M., Wessler, S., Tegtmeyer, N., 2013. Transmigration route of *Campylobacter jejuni* across polarized intestinal epithelial cells: paracellular, transcellular or both? Cell Commun. Signal. 11, 72.

Baek, K.T., Vegge, C.S., Brondsted, L., 2011. HtrA chaperone activity contributes to host cell binding in *Campylobacter jejuni*. Gut. Pathog. 3, 13.

Barrero-Tobon, A.M., Hendrixson, D.R., 2012. Identification and analysis of flagellar coexpressed determinants (Feds) of *Campylobacter jejuni* involved in colonization. Mol. Microbiol. 84, 352–369.

Barrero-Tobon, A.M., Hendrixson, D.R., 2014. Flagellar biosynthesis exerts temporal regulation of secretion of specific *Campylobacter jejuni* colonization and virulence determinants. Mol. Microbiol. 93, 957–974.

Bereswill, S., Plickert, R., Fischer, A., Kühl, A.A., Loddenkemper, C., Batra, A., Siegmund, B., Göbel, U.B., Heimesaat, M.M., 2011. What you eat is what you get: novel *Campylobacter* models in the quadrangle relationship between nutrition, obesity, microbiota and susceptibility to infection. Eur. J. Microbiol. Immunol. (Bp) 1, 237–248.

Bickerstaff, E.R., 1957. Brain-stem encephalitis; further observations on a grave syndrome with benign prognosis. Br. Med. J. 1, 1384–1387.

Blaser, M.J., Hoverson, D., Ely, I.G., Duncan, D.J., Wang, W.L., Brown, W.R., 1984. Studies of *Campylobacter jejuni* in patients with inflammatory bowel disease. Gastroenterology 86, 33–38.

Boehm, M., Krause-Gruszczynska, M., Tegtmeyer, N., Takahashi, S., Backert, S., 2011. Role of fibronectin, integrin-β1, FAK, Tiam1, DOCK180 in activating Rho GTPase Rac1. Front. Cell Infect. Microbiol. 1, 17.

Boehm, M., Hoy, B., Rohde, M., Tegtmeyer, N., Bæk, K.T., Oyarzabal, O.A., Brøndsted, L., Wessler, S., Backert, S., 2012. Rapid paracellular transmigration of *C. jejuni* across polarized epithelial cells without affecting TER: role of proteolytic-active HtrA cleaving E-cadherin but not fibronectin. Gut Pathog. 4, 3.

Boehm, M., Lind, J., Backert, S., Tegtmeyer, N., 2015. *Campylobacter jejuni* serine protease HtrA plays an important role in heat tolerance, oxygen resistance, host cell adhesion, invasion, and transmigration. Eur. J. Microbiol. Immunol. (Bp) 5, 68–80.

Boyanova, L., Gergova, G., Spassova, Z., Koumanova, R., Yaneva, P., Mitov, I., Derejian, S., Krastev, Z., 2004. *Campylobacter* infection in 682 Bulgarian patients with acute enterocolitis, inflammatory bowel disease, and other chronic intestinal diseases. Diagn. Microbiol. Infect. Dis. 49, 71–74.

Braun, J., Yin, Z., Spiller, I., Siegert, S., Rudwaleit, M., Liu, L., Radbruch, A., Sieper, J., 1999. Low secretion of TNFα, but no other Th1 or Th2 cytokines, by PBMC correlates with chronicity in reactive arthritis. Arthritis Rheum. 42, 2039–2044.

Brøndsted, L., Andersen, M.T., Parker, M., Jørgensen, K., Ingmer, H., 2005. HtrA protease of *C.jejuni* is required for heat and oxygen tolerance and for optimal interaction with epithelial cells. Appl. Environ. Microbiol. 71, 3205–3212.

Bronowski, C., James, C.E., Winstanley, C., 2014. Role of environmental survival in transmission of *Campylobacter jejuni*. FEMS Microbiol. Lett. 356, 8–19.

Butrimiene, I., Jarmalaite, S., Ranceva, J., Venalis, A., Jasiuleviciute, L., Zvirbliene, A., 2004. Different cytokine profiles in patients with chronic and acute reactive arthritis. Rheumatology (Oxford) 43, 1300–1304.

Calin, A., Fries, J.F., 1976. An "experimental" epidemic of Reiter's syndrome revisited. Follow-up evidence on genetic and environmental factors. Ann. Intern. Med. 84, 564–566.

Carter, J.D., 2006. Reactive arthritis: defined etiologies, emerging pathophysiology, and unresolved treatment. Infect. Dis. Clin. North Am. 20, 827–847.

Castano-Rodriguez, N., Kaakoush, N.O., Lee, W.S., Mitchell, H.M., 2015. Dual role of *Helicobacter* and *Campylobacter* species in IBD: a systematic review and meta-analysis. Gut.

Cherrington, C.A., Hinton, M., Pearson, G.R., Chopra, I., 1991. Short-chain organic acids at pH 5.0 kill *Escherichia coli* and *Salmonella* spp. without causing membrane perturbation. J. Appl. Bacteriol. 70, 161–165.

Christensen, J.E., Pacheco, S.A., Konkel, M.E., 2009. Identification of a *Campylobacter jejuni*—secreted protein required for maximal invasion of host cells. Mol. Microbiol. 73, 650–662.

Crim, S.M., Griffin, P.M., Tauxe, R., Marder, E.P., Gilliss, D., Cronquist, A.B., Cartter, M., Tobin-D'Angelo, M., Blythe, D., Smith, K., Lathrop, S., Zansky, S., Cieslak, P.R., Dunn, J., Holt, K.G., Wolpert, B., Henao, O.L., 2015. Preliminary incidence and trends of infection with pathogens transmitted commonly through food - Foodborne Diseases Active Surveillance Network, 10 U.S. sites, 2006–2014. Morb. Mortal. Wkly. Rep. 64, 495–499.

Dasti, J.I., Tareen, A.M., Lugert, R., Zautner, A.E., Gross, U., 2010. *Campylobacter jejuni*: a brief overview on pathogenicity-associated factors and disease-mediating mechanisms. Int. J. Med. Microbiol. 300, 205–211.

Dunlop, S.P., Jenkins, D., Neal, K.R., Spiller, R.C., 2003. Relative importance of enterochromaffin cell hyperplasia, anxiety, and depression in postinfectious IBS. Gastroenterology 125, 1651–1659.

Eckburg, P.B., Bik, E.M., Bernstein, C.N., Purdom, E., Dethlefsen, L., Sargent, M., Gill, S.R., Nelson, K.E., Relman, D.A., 2005. Microbiology: diversity of the human intestinal microbial flora. Science 308, 1635–1638.

Ellis-Iversen, J., Jorgensen, F., Bull, S., Powell, L., Cook, A.J., Humphrey, T.J., 2009. Risk factors for *Campylobacter* colonisation during rearing of broiler flocks in Great Britain. Prev. Vet. Med. 89, 178–184.

Elmi, A., Nasher, F., Jagatia, H., Gundogdu, O., Bajaj-Elliott, M., Wren, B.W., Dorrell, N., 2015. *Campylobacter jejuni* outer membrane vesicle-associated proteolytic activity promotes bacterial invasion by mediating cleavage of intestinal epithelial cell E-cadherin and occludin. Cell Microbiol. 18, 561–572.

Engberg, J., On, S.L., Harrington, C.S., Gerner-Smidt, P., 2000. Prevalence of *Campylobacter*, *Arcobacter*, *Helicobacter*, and *Sutterella* spp. in human fecal samples as estimated by a reevaluation of isolation methods for Campylobacters. J. Clin. Microbiol. 38, 286–291.

Eucker, T.P., Konkel, M.E., 2012. The cooperative action of bacterial fibronectin-binding proteins and secreted proteins promote maximal *C. jejuni* invasion of host cells by stimulating membrane ruffling. Cell Microbiol. 14, 226–238.

Farmer, S., Keenan, A., Vivancos, R., 2012. Food-borne *Campylobacter* outbreak in Liverpool associated with cross-contamination from chicken liver parfait: implications for investigation of similar outbreaks. Public Health 126, 657–659.

Fisher, M., 1956. An unusual variant of acute idiopathic polyneuritis (syndrome of ophthalmoplegia, ataxia and areflexia). N. Engl. J. Med. 255, 57–65.

Fouts, D.E., Mongodin, E.F., Mandrell, R.E., Miller, W.G., Rasko, D.A., Ravel, J., Brinkac, L.M., DeBoy, R.T., Parker, C.T., Daugherty, S.C., Dodson, R.J., Durkin, A.S., Madupu, R., Sullivan, S.A., Shetty, J.U., Ayodeji, M.A., Shvartsbeyn, A., Schatz, M.C., Badger, J.H., Fraser, C.M., Nelson, K.E., 2005. Major structural differences and novel potential virulence mechanisms from the genomes of multiple *Campylobacter* species. PLoS Biol. 3, e15.

Geleijns, K., Roos, A., Houwing-Duistermaat, J.J., van Rijs, W., Tio-Gillen, A.P., Laman, J.D., van Doorn, P.A., Jacobs, B.C., 2006. Mannose-binding lectin contributes to the severity of GBS. J. Immunol. 177, 4211–4217.

Godschalk, P.C., Heikema, B.C., Gilbert, M., Komagamine, T., Ang, C.W., Glerum, J., Brochu, D., Li, J., Yuki, N., Jacobs, B.C., van Belkum, A., Endtz, H.P., 2004. The crucial role of *Campylobacter jejuni* genes in anti-ganglioside antibody induction in Guillain-Barre syndrome. J. Clin. Invest. 114, 1659–1665.

Gormley, F.J., Little, C.L., Rawal, N., Gillespie, I.A., Lebaigue, S., Adak, G.K., 2011. A 17-year Review of Foodborne outbreaks: describing the continuing decline in England and Wales (1992–2008). Epidemiol. Infect. 139, 688–699.

Gould, L.H., Walsh, K.A., Vieira, A.R., Herman, K., Williams, I.T., Hall, A.J., Cole, D., 2013. Surveillance for foodborne disease outbreaks—United States, 1998–2008. Morb. Mortal. Wkly. Rep. 62, 1–34.

Gradel, K.O., Nielsen, H.L., Schonheyder, H.C., Ejlertsen, T., Kristensen, B., Nielsen, H., 2009. Increased short- and long-term risk of inflammatory bowel disease after *Salmonella* or *Campylobacter* gastroenteritis. Gastroenterology 137, 495–501.

Grover, M., 2014. Role of gut pathogens in development of irritable bowel syndrome. Indian J. Med. Res. 139, 11–18.

Guccione, E., Hitchcock, A., Hall, S.J., Mulholland, F., Shearer, N., van Vliet, A.H., Kelly, D.J., 2010. Reduction of fumarate, mesaconate and crotonate by Mfr, a novel periplasmic reductase in *C. jejuni*. Environ. Microbiol. 12, 576–591.

Guillain, G., Barré, J.A., Strohl, A., 1916. Radiculoneuritis syndrome with hyperalbuminosis of cerebrospinal fluid without cellular reaction. Notes on clinical features and graphs of tendon reflexes. Ann. Med. Interne 150, 24–32.

Guzman-Herrador, B., Carlander, A., Ethelberg, S., Freiesleben de Blasio, B., Kuusi, M., Lund, V., Löfdahl, M., MacDonald, E., Nichols, G., Schönning, C., Sudre, B., Trönnberg, L., Vold, L., Semenza, J.C., Nygård, K., 2015. Waterborne outbreaks in the Nordic countries, 1998–2012. Euro Surveill. 20, 21160.

Haag, L.M., Fischer, A., Otto, B., Plickert, R., Kühl, A.A., Göbel, U.B., Bereswill, S., Heimesaat, M.M., 2012. Intestinal microbiota shifts towards elevated commensal *Escherichia coli* loads abrogate colonization resistance against *Campylobacter jejuni* in mice. PLoS One 7, e35988.

Halstead, S.K., Zitman, F.M., Humphreys, P.D., Greenshields, K., Verschuuren, J.J., Jacobs, B.C., Rother, R.P., Plomp, J.J., Willison, H.J., 2008. Eculizumab prevents anti-ganglioside antibody-mediated neuropathy in a murine model. Brain 131, 1197–1208.

Hannu, T., Mattila, L., Rautelin, H., Pelkonen, P., Lahdenne, P., Siitonen, A., Leirisalo-Repo, M., 2002. *Campylobacter*-triggered reactive arthritis: a population-based study. Rheumatology (Oxford) 41, 312–318.

Hauri, A.M., Just, M., McFarland, S., Schweigmann, A., Schlez, K., Krahn, J., 2013. [Campy-lobacteriosis outbreaks in the state of Hesse, Germany, 2005–2011: raw milk yet again]. Dtsch. Med. Wochenschr. 138, 357–361.

Havelaar, A.H., van Pelt, W., Ang, C.W., Wagenaar, J.A., van Putten, J.P., Gross, U., Newell, D.G., 2009. Immunity to *Campylobacter*: its role in risk assessment and epidemiology. Crit. Rev. Microbiol. 35, 1–22.

Heikema, A.P., Jacobs, B.C., Horst-Kreft, D., Huizinga, R., Kuijf, M.L., Endtz, H.P., Samsom, J.N., van Wamel, W.J., 2013. Siglec-7 specifically recognizes *C. jejuni* strains associated with oculomotor weakness in Guillain-Barre syndrome and Miller Fisher syndrome. Clin. Microbiol. Infect. 19, E106–E112.

Heikema, A.P., Islam, Z., Horst-Kreft, D., Huizinga, R., Jacobs, B.C., Wagenaar, J.A., Poly, F., Guerry, P., van Belkum, A., Parker, C.T., Endtz, H.P., 2015. *Campylobacter jejuni* capsular geno-types are related to Guillain-Barre syndrome. Clin. Microbiol. Infect. 21, 852, e851–e859.

Herman, L., Heyndrickx, M., Grijspeerdt, K., Vandekerchove, D., Rollier, I., De Zutter, L., 2003. Routes for *Campylobacter* contamination of poultry meat: epidemiological study from hatchery to slaughter. Epidemiol. Infect. 131, 1169–1180.

Hill Gaston, J.S., Lillicrap, M.S., 2003. Arthritis associated with enteric infection. Best Pract. Res. Clin. Rheumatol. 17, 219–239.

Hofreuter, D., 2014. Defining the metabolic requirements for the growth and colonization capacity of *C. jejuni*. Front. Cell Infect. Microbiol. 4, 137.

Hofreuter, D., Tsai, J., Watson, R.O., Novik, V., Altman, B., Benitez, M., Clark, C., Perbost, C., Jarvie, T., Du, L., Galán, J.E., 2006. Unique features of a highly pathogenic *Campylo-bacter jejuni* strain. Infect. Immun. 74, 4694–4707.

Hoy, B., Löwer, M., Weydig, C., Carra, G., Tegtmeyer, N., Geppert, T., Schröder, P., Sewald, N., Backert, S., Schneider, G., Wessler, S., 2010. *Helicobacter pylori* HtrA is a new se-creted virulence factor that cleaves E-cadherin to disrupt intercellular adhesion. EMBO Rep. 11, 798–804.

Hoy, B., Geppert, T., Boehm, M., Reisen, F., Plattner, P., Gadermaier, G., Sewald, N., Ferreira, F., Briza, P., Schneider, G., Backert, S., Wessler, S., 2012. Distinct roles of secreted HtrA proteases from Gram-negative pathogens in cleaving the junctional protein and tumor sup-pressor E-cadherin. J. Biol. Chem. 287, 10115–10120.

Hughes, R.A., Cornblath, D.R., 2005. Guillain-Barre syndrome. Lancet 366, 1653–1666.

Hughes, R.A., Swan, A.V., Raphael, J.C., Annane, D., van Koningsveld, R., van Doorn, P.A., 2007. Immunotherapy for Guillain-Barre syndrome: a systematic review. Brain 130, 2245–2257.

Huizinga, R., van Rijs, W., Bajramovic, J.J., Kuijf, M.L., Laman, J.D., Samsom, J.N., Jacobs, B.C., 2013. Sialylation of *Campylobacter jejuni* endotoxin promotes dendritic cell-medi-ated B cell responses through CD14-dependent production of IFN-beta and TNF-alpha. J. Immunol. 191, 5636–5645.

Huizinga, R., van den Berg, B., van Rijs, W., Tio-Gillen, A.P., Fokkink, W.J., Bakker-Jonges, L.E., Geleijns, K., Samsom, J.N., van Doorn, P.A., Jacobs, B.C., 2015. Innate immunity to *C. jejuni* in GBS. Ann. Neurol. 78, 343–354.

Islam, Z., Jacobs, B.C., van Belkum, A., Mohammad, Q.D., Islam, M.B., Herbrink, P., Diordit-sa, S., Luby, S.P., Talukder, K.A., Endtz, H.P., 2010. Axonal variant of Guillain-Barre syn-drome associated with *Campylobacter* infection in Bangladesh. Neurology 74, 581–587.

Jackson, B.R., Zegarra, J.A., Lopez-Gatell, H., Sejvar, J., Arzate, F., Waterman, S., Nunez, A.S., Lopez, B., Weiss, J., Cruz, R.Q., Murrieta, D.Y., Luna-Gierke, R., Heiman, K., Vieira, A.R., Fitzgerald, C., Kwan, P., Zarate-Bermudez, M., Talkington, D., Hill, V.R.,

Mahon, B., Team, G.B.S.O.I., 2014. Binational outbreak of Guillain-Barre syndrome associated with *Campylobacter jejuni* infection, Mexico and USA, 2011. Epidemiol. Infect. 142, 1089–1099.

Jacobs, B.C., Rothbarth, P.H., van der Meche, F.G., Herbrink, P., Schmitz, P.I., de Klerk, M.A., van Doorn, P.A., 1998. The spectrum of antecedent infections in Guillain-Barre syndrome: a case-control study. Neurology 51, 1110–1115.

Jin, S., Joe, A., Lynett, J., Hani, E.K., Sherman, P., Chan, V.L., 2001. JlpA, a novel surface-exposed lipoprotein specific to *Campylobacter jejuni*, mediates adherence to host epithelial cells. Mol. Microbiol. 39, 1225–1236.

Jin, S., Song, Y.C., Emili, A., Sherman, P.M., Chan, V.L., 2003. JlpA of *Campylobacter jejuni* interacts with surface-exposed heat shock protein 90alpha and triggers signalling pathways leading to the activation of NF-kappaB and p38 MAP kinase in epithelial cells. Cell Microbiol. 5, 165–174.

Jorgensen, F., Ellis-Iversen, J., Rushton, S., Bull, S.A., Harris, S.A., Bryan, S.J., Gonzalez, A., Humphrey, T.J., 2011. Influence of season and geography on *Campylobacter jejuni* and *C. coli* subtypes in housed broiler flocks reared in Great Britain. Appl. Environ. Microbiol. 77, 3741–3748.

Kaakoush, N.O., Castano-Rodriguez, N., Mitchell, H.M., Man, S.M., 2015. Global epidemiology of *Campylobacter* infection. Clin. Microbiol. Rev. 28, 687–720.

Kaluza, W., Leirisalo-Repo, M., Marker-Hermann, E., Westman, P., Reuss, E., Hug, R., Mastrovic, K., Stradmann-Bellinghausen, B., Granfors, K., Galle, P.R., Hohler, T., 2001. IL10.G microsatellites mark promoter haplotypes associated with protection against the development of reactive arthritis in Finnish patients. Arthritis Rheum. 44, 1209–1214.

Keithlin, J., Sargeant, J., Thomas, M.K., Fazi, A., 2014. Systematic review and meta-analysis of the proportion of *Campylobacter* cases that develop chronic sequelae. BMC Public Health 14, 1203.

Kelle, K., Pagés, J.M., Bolla, J.M., 1998. A putative adhesin gene cloned from *Campylobacter jejuni*. Res. Microbiol. 149, 723–733.

Konkel, M.E., Garvis, S.G., Tipton, S.L., Anderson, Jr., D.E., Cieplak, Jr., W., 1997. Identification and molecular cloning of a gene encoding a fibronectin-binding protein (CadF) from *Campylobacter jejuni*. Mol. Microbiol. 24, 953–963.

Konkel, M.E., Kim, B.J., Rivera-Amill, V., Garvis, S.G., 1999. Bacterial secreted proteins are required for the internalization of *Campylobacter jejuni* into cultured mammalian cells. Mol. Microbiol. 32, 691–701.

Konkel, M.E., Christensen, J.E., Keech, A.M., Monteville, M.R., Klena, J.D., Garvis, S.G., 2005. Identification of a fibronectin-binding domain within the *Campylobacter jejuni* CadF protein. Mol. Microbiol. 57, 1022–1035.

Konkel, M.E., Larson, C.L., Flanagan, R.C., 2010. *C. jejuni* FlpA binds fibronectin and is required for maximal host cell adherence. J. Bacteriol. 192, 68–76.

Krause-Gruszczynska, M., Rohde, M., Hartig, R., Genth, H., Schmidt, G., Keo, T., König, W., Miller, W.G., Konkel, M.E., Backert, S., 2007. Role of the small Rho GTPases Rac1 and Cdc42 in host cell invasion of *Campylobacter jejuni*. Cell Microbiol. 9, 2431–2444.

Krause-Gruszczynska, M., Boehm, M., Rohde, M., Tegtmeyer, N., Takahashi, S., Buday, L., Oyarzabal, O.A., Backert, S., 2011. The signaling pathway of *C. jejuni* induced Cdc42 activation: Role of fibronectin integrin beta1, tyrosine kinases and guanine exchange factor Vav2. Cell Commun. Signal. 9, 32.

Kuijf, M.L., Samsom, J.N., van Rijs, W., Bax, M., Huizinga, R., Heikema, A.P., van Doorn, P.A., van Belkum, A., van Kooyk, Y., Burgers, P.C., Luider, T.M., Endtz, H.P.,

Nieuwenhuis, E.E., Jacobs, B.C., 2010. TLR4-mediated sensing of *Campylobacter jejuni* by dendritic cells is determined by sialylation. J. Immunol. 185, 748–755.

Kwan, P.S.L., Barrigas, M., Bolton, F.J., French, N.P., Gowland, P., Kemp, R., Leatherbarrow, H., Upton, M., Fox, A.J., 2008. Molecular epidemiology of *Campylobacter jejuni* populations in dairy cattle, wildlife, and the environment in a farmland area. Appl. Environ. Microbiol. 74, 5130–5138.

Leirisalo, M., Skylv, G., Kousa, M., Voipio-Pulkki, L.M., Suoranta, H., Nissila, M., Hvidman, L., Nielsen, E.D., Svejgaard, A., Tilikainen, A., Laitinen, O., 1982. Followup study on patients with Reiter's disease and reactive arthritis, with special reference to HLA-B27. Arthritis Rheum. 25, 249–259.

Macfarlane, S., Macfarlane, G.T., 2003. Regulation of short-chain fatty acid production. Proc. Nutr. Soc. 62, 67–72.

Mahendran, V., Riordan, S.M., Grimm, M.C., Tran, T.A., Major, J., Kaakoush, N.O., Mitchell, H., Zhang, L., 2011. Prevalence of *Campylobacter* species in adult Crohn's disease and the preferential colonization sites of *Campylobacter* species in the human intestine. PLoS One 6, e25417.

Mahendran, V., Tan, Y.S., Riordan, S.M., Grimm, M.C., Day, A.S., Lemberg, D.A., Octavia, S., Zhang, L., 2013. The prevalence and polymorphisms of zonula occluden toxin gene in multiple *C. concisus* strains isolated from saliva of patients with inflammatory bowel disease and controls. PLoS One 8, e75525.

Marshall, J.K., Thabane, M., Borgaonkar, M.R., James, C., 2007. Postinfectious irritable bowel syndrome after a food-borne outbreak of acute gastroenteritis attributed to a viral pathogen. Clin. Gastroenterol. Hepatol. 5, 457–460.

Marteyn, B., Scorza, F.B., Sansonetti, P.J., Tang, C., 2011. Breathing life into pathogens: the influence of oxygen on bacterial virulence and host responses in the gastrointestinal tract. Cell Microbiol. 13, 171–176.

Masanta, W.O., Heimesaat, M.M., Bereswill, S., Tareen, A.M., Lugert, R., Groß, U., Zautner, A.E., 2013. Modification of intestinal microbiota and its consequences for innate immune response in the pathogenesis of campylobacteriosis. Clin. Dev. Immunol. 2013, 526860.

McFaydean, J., Stockman, S., 1913. Report of the Departmental Committee appointed by the board of agriculture and fisheries to inquire into epizoonotic abortion. Appendix to part III: Abortion in sheep. His majesty's stationery office, London, UK, 1-29.

Molodecky, N.A., Soon, I.S., Rabi, D.M., Ghali, W.A., Ferris, M., Chernoff, G., Benchimol, E.I., Panaccione, R., Ghosh, S., Barkema, H.W., Kaplan, G.G., 2012. Increasing incidence and prevalence of the inflammatory bowel diseases with time, based on systematic review. Gastroenterology 142, 46–54.

Monteville, M.R., Konkel, M.E., 2002. Fibronectin-facilitated invasion of T84 eukaryotic cells by *Campylobacter jejuni* occurs preferentially at the basolateral cell surface. Infect. Immun. 70, 6665–6671.

Monteville, M.R., Yoon, J.E., Konkel, M.E., 2003. Maximal adherence and invasion of INT407 cells by *Campylobacter jejuni* requires the CadF outer-membrane protein and microfilament reorganization. Microbiology 149, 153–165.

Moser, I., Schroeder, W., Salnikow, J., 1997. *Campylobacter jejuni* major outer membrane protein and a 59-kDa protein are involved in binding to fibronectin and INT-407 cell membranes. FEMS Microbiol. Lett. 157, 233–238.

Nachamkin, I., Allos, B.M., Ho, T., 1998. *Campylobacter* species and Guillain-Barre syndrome. Clin. Microbiol. Rev. 11, 555–567.

Nachamkin, I., Szymanski, C.M., Blaser, M.J., 2008. *Campylobacter*, third ed. ASM Press, Washington, DC.

Newell, D.G., McBride, H., Pearson, A.D., 1984. The identification of outer membrane proteins and flagella of *C. jejuni*. J. Gen. Microbiol. 130, 1201–1208.

Novik, V., Hofreuter, D., Galan, J.E., 2010. Identification of *C. jejuni* genes involved in its interaction with epithelial cells. Infect. Immun. 78, 3540–3553.

Ó Cróinín, T., Backert, S., 2012. Host epithelial cell invasion by *Campylobacter jejuni*: trigger or zipper mechanism? Front. Cell Infect. Microbiol. 2, 25.

Oelschlaeger, T.A., Guerry, P., Kopecko, D.J., 1993. Unusual microtubule-dependent endocytosis mechanisms triggered by *Campylobacter jejuni* and *Citrobacter freundii*. Proc. Natl. Acad. Sci. USA 90, 6884–6888.

Parkhill, J., Wren, B.W., Mungall, K., Ketley, J.M., Churcher, C., Basham, D., Chillingworth, T., Davies, R.M., Feltwell, T., Holroyd, S., Jagels, K., Karlyshev, A.V., Moule, S., Pallen, M.J., Penn, C.W., Quail, M.A., Rajandream, M.A., Rutherford, K.M., van Vliet, A.H., Whitehead, S., Barrell, B.G., 2000. The genome sequence of the food-borne pathogen *Campylobacter jejuni* reveals hypervariable sequences. Nature 403, 665–668.

Pei, Z., Burucoa, C., Grignon, B., Baqar, S., Huang, X.Z., Kopecko, D.J., Bourgeois, A.L., Fauchere, J.L., Blaser, M.J., 1998. Mutation in the *peb1A* locus of *Campylobacter jejuni* reduces interactions with epithelial cells and intestinal colonization of mice. Infect. Immun. 66, 938–944.

Platts-Mills, J.A., Kosek, M., 2014. Update on the burden of *Campylobacter* in developing countries. Curr. Opin. Infect. Dis. 27, 444–450.

Poly, F., Guerry, P., 2008. Pathogenesis of *Campylobacter*. Curr. Opin. Gastroenterol. 24, 27–31.

Pope, J.E., Krizova, A., Garg, A.X., Thiessen-Philbrook, H., Ouimet, J.M., 2007. *Campylobacter* reactive arthritis: a systematic review. Semin. Arthritis. Rheum. 37, 48–55.

Potturi-Venkata, L.P., Backert, S., Vieira, S.L., Oyarzabal, O.A., 2007. Evaluation of logistic processing to reduce cross-contamination of commercial broiler carcasses with *Campylobacter* spp. J. Food Prot. 70, 2549–2554.

Rinaldi, S., Brennan, K.M., Kalna, G., Walgaard, C., van Doorn, P., Jacobs, B.C., Yu, R.K., Mansson, J.E., Goodyear, C.S., Willison, H.J., 2013. Antibodies to heteromeric glycolipid complexes in GBS. PLoS One 8, e82337.

Rivera-Amill, V., Kim, B.J., Seshu, J., Konkel, M.E., 2001. Secretion of the virulence-associated *Campylobacter* invasion antigens from *Campylobacter jejuni* requires a stimulatory signal. J. Infect. Dis. 183, 1607–1616.

Sartor, R.B., 1995. Current concepts of the etiology and pathogenesis of ulcerative colitis and Crohn's disease. Gastroenterol. Clin. North Am. 24, 475–507.

Schmidt-Ott, R., Schmidt, H., Feldmann, S., Brass, F., Krone, B., Gross, U., 2006. Improved serological diagnosis stresses the major role of *Campylobacter jejuni* in triggering GBS. Clin. Vaccine Immunol. 13, 779–783.

Scott, M.K., Geissler, A., Poissant, T., DeBess, E., Melius, B., Eckmann, K., Salehi, E., Cieslak, P.R., 2015. Notes from the field: campylobacteriosis outbreak associated with consuming undercooked chicken liver pâté - Ohio and Oregon, December 2013–January 2014. Morb. Mortal. Wkly. Rep. 64, 399.

Sheppard, S.K., Maiden, M.C., 2015. The evolution of *Campylobacter jejuni* and *Campylobacter coli*. Cold Spring Harb. Perspect. Biol. 7 (8), a018119, 22.

Siamer, S., Dehio, C., 2015. New insights into the role of *Bartonella* effector proteins in pathogenesis. Curr. Opin. Microbiol. 23, 80–85.

Song, Y.C., Jin, S., Louie, H., Ng, D., Lau, R., Zhang, Y., Weerasekera, R., Al Rashid, S., Ward, L.A., Der, S.D., Chan, V.L., 2004. FlaC, a protein of *Campylobacter jejuni* TGH9011 secreted through the flagellar apparatus, binds epithelial cells and influences cell invasion. Mol. Microbiol. 53, 541–553.

Spiller, R., Garsed, K., 2009. Postinfectious irritable bowel syndrome. Gastroenterology 136, 1979–1988.

Stingl, K., Knüver, M.T., Vogt, P., Buhler, C., Krüger, N.J., Tenhagen, B.A., Hartung, M., Schroeter, A., Ellerbroek, L., Appel, B., Käsbohrer, A., 2012. Quo vadis? Monitoring *Campylobacter* in Germany. Eur. J. Microbiol. Immunol. 2, 88–96.

Szymanski, C.M., King, M., Haardt, M., Armstrong, G.D., 1995. *Campylobacter jejuni* motility and invasion of Caco-2 cells. Infect. Immun. 63, 4295–4300.

Tam, C.C., Rodrigues, L.C., Petersen, I., Islam, A., Hayward, A., O'Brien, S.J., 2006. Incidence of GBS among patients with *Campylobacter* infection: a general practice research database study. J. Infect. Dis. 194, 95–97.

Tam, C.C., Rodrigues, L.C., Viviani, L., Dodds, J.P., Evans, M.R., Hunter, P.R., Gray, J.J., Letley, L.H., Rait, G., Tompkins, D.S., O'Brien, S.J., 2012. Longitudinal study of infectious intestinal disease in the UK: incidence in the community and presenting to general practice. Gut 61, 69–77.

Thomas, M.T., Shepherd, M., Poole, R.K., van Vliet, A.H., Kelly, D.J., Pearson, B.M., 2011. Two respiratory enzyme systems in *Campylobacter jejuni* NCTC11168 contribute to growth on L-lactate. Environ. Microbiol. 13, 48–61.

Thornley, J.P., Jenkins, D., Neal, K., Wright, T., Brough, J., Spiller, R.C., 2001. Relationship of *Campylobacter* toxigenicity in vitro to the development of postinfectious irritable bowel syndrome. J. Infect. Dis. 184, 606–609.

Tindberg, Y., Bengtsson, C., Bergstrom, M., Granstrom, M., 2001. The accuracy of serologic diagnosis of *Helicobacter pylori* infection in school-aged children of mixed ethnicity. Helicobacter 6, 24–30.

Townes, J.M., 2010. Reactive arthritis after enteric infections in the United States: the problem of definition. Clin. Infect. Dis. 50, 247–254.

van den Berg, B., Walgaard, C., Drenthen, J., Fokke, C., Jacobs, B.C., van Doorn, P.A., 2014. Guillain-Barre syndrome: pathogenesis, diagnosis, treatment and prognosis. Nat. Rev. Neurol. 10, 469–482.

van Gerwe, T., Miflin, J.K., Templeton, J.M., Bouma, A., Wagenaar, J.A., Jacobs-Reitsma, W.F., Stegeman, A., Klinkenberg, D., 2009. Quantifying transmission of *C. jejuni* in commercial broiler flocks. Appl. Environ. Microbiol. 75, 625–628.

van Spreeuwel, J.P., Duursma, G.C., Meijer, C.J., Bax, R., Rosekrans, P.C., Lindeman, J., 1985. *Campylobacter colitis*: histological, immunohistochemical and ultrastructural findings. Gut 26, 945–951.

Verhoeff-Bakkenes, L., Jansen, H.A.P.M., in't Veld, P.H., Beumer, R.R., 2011. Consumption of raw vegetables and fruits: a risk factor for *Campylobacter* infections. Int. J. Food Microbiol. 144, 406–412.

Wakerley, B.R., Yuki, N., 2015. Guillain-Barre syndrome. Expert Rev. Neurother. 15, 847–849.

Wassenaar, T.M., Blaser, M.J., 1999. Pathophysiology of *Campylobacter jejuni* infections of humans. Microbes Infect. 1, 1023–1033.

Weber, P., Koch, M., Heizmann, W.R., Scheurlen, M., Jenss, H., Hartmann, F., 1992. Microbic superinfection in relapse of inflammatory bowel disease. J. Clin. Gastroenterol. 14, 302–308.

Wedderkopp, A., Nielsen, E.M., Pedersen, K., 2003. Distribution of *C. jejuni* Penner serotypes in broiler flocks 1998–2000 in a small Danish community with special reference to serotype 4-complex. Epidemiol. Infect. 131, 915–921.

Weerakoon, D.R., Borden, N.J., Goodson, C.M., Grimes, J., Olson, J.W., 2009. The role of respiratory donor enzymes in *Campylobacter jejuni* host colonization and physiology. Microb. Pathog. 47, 8–15.

World Health Organization, 2014. Foodborne diseases. WHO, Geneva, Switzerland. Available from: http://www.who.int/foodsafety/areas_work/foodborne-diseases/en/

Wright, J.A., Grant, A.J., Hurd, D., Harrison, M., Guccione, E.J., Kelly, D.J., Maskell, D.J., 2009. Metabolite and transcriptome analysis of *C. jejuni* in vitro growth reveals a stationary-phase physiological switch. Microbiology 155, 80–94.

Wu, I.B., Schwartz, R.A., 2008. Reiter's syndrome: the classic triad and more. J. Am. Acad. Dermatol. 59, 113–121.

Wu, L.Y., Zhou, Y., Qin, C., Hu, B.L., 2012. The effect of TNF-alpha, FcγR and CD1 polymorphisms on Guillain-Barre syndrome risk: evidences from a meta-analysis. J. Neuroimmunol. 243, 18–24.

Yoon, S.S., Kim, E.K., Lee, W.J., 2015. Functional genomic and metagenomic approaches to understanding gut microbiota-animal mutualism. Curr. Opin. Microbiol. 24, 38–46.

Young, G.M., Schmiel, D.H., Miller, V.L., 1999. A new pathway for the secretion of virulence factors by bacteria: the flagellar export apparatus functions as a protein-secretion system. Proc. Natl. Acad. Sci. USA 96, 6456–6461.

Young, K.T., Davis, L.M., Di Rita, V.J., 2007. *Campylobacter jejuni*: molecular biology and pathogenesis. Nat. Rev. Microbiol. 5, 665–679.

Yuki, N., Hartung, H.P., 2012. Guillain-Barre syndrome. N. Engl. J. Med. 366, 2294–2304.

Yuki, N., Koga, M., 2006. Bacterial infections in Guillain-Barre and Fisher syndromes. Curr. Opin. Neurol. 19, 451–457.

Yuki, N., Susuki, K., Koga, M., Nishimoto, Y., Odaka, M., Hirata, K., Taguchi, K., Miyatake, T., Furukawa, K., Kobata, T., Yamada, M., 2004. Carbohydrate mimicry between human ganglioside GM1 and *C. jejuni* lipooligosaccharide causes GBS. Proc. Natl. Acad. Sci. USA 101, 11404–11409.

Zautner, A.E., Johann, C., Strubel, A., Busse, C., Tareen, A.M., Masanta, W.O., Lugert, R., Schmidt-Ott, R., Gross, U., 2014. Seroprevalence of campylobacteriosis and relevant post-infectious sequelae. Eur. J. Clin. Microbiol. Infect. Dis. 33, 1019–1027.

Zhang, L., 2015. Oral *Campylobacter* species: initiators of a subgroup of inflammatory bowel disease? World J. Gastroenterol. 21, 9239–9244.

Zhang, L., Man, S.M., Day, A.S., Leach, S.T., Lemberg, D.A., Dutt, S., Stormon, M., Otley, A., O'Loughlin, E.V., Magoffin, A., Ng, P.H., Mitchell, H., 2009. Detection and isolation of *Campylobacter* species other than *C. jejuni* from children with Crohn's disease. J. Clin. Microbiol. 47, 453–455.

Zhang, L., Lee, H., Grimm, M.C., Riordan, S.M., Day, A.S., Lemberg, D.A., 2014. *Campylobacter concisus* and inflammatory bowel disease. World J. Gastroenterol. 20, 1259–1267.

Ziprin, R.L., Young, C.R., Stanker, L.H., Hume, M.E., Konkel, M.E., 1999. The absence of cecal colonization of chicks by a mutant of *Campylobacter jejuni* not expressing bacterial fibronectin binding protein. Avian Dis. 43, 586–589.

Ziprin, R.L., Young, C.R., Byrd, J.A., Stanker, L.H., Hume, M.E., Gray, S.A., Kim, B.J., Konkel, M.E., 2001. Role of *Campylobacter jejuni* potential virulence genes in cecal colonization. Avian Dis. 45, 549–557.

Health and economic burden of *Campylobacter*

Brecht Devleesschauwer*, Martijn Bouwknegt,
Marie-Josée J. Mangen**,†, Arie H. Havelaar‡**

**Department of Public Health and Surveillance, Scientific Institute of Public Health (WIV-ISP),
Brussels, Belgium; **Centre for Infectious Disease Control, National Institute for Public
Health and the Environment (RIVM), Bilthoven, The Netherlands; †Department
of Public Health, Health Technology Assessment and Medical Humanities, Julius Center
for Health Sciences and Primary Care, University Medical Center Utrecht, Utrecht,
The Netherlands; ‡Emerging Pathogens Institute, Department of Animal Sciences and
Institute for Sustainable Food Systems, University of Florida, Gainesville, FL, United States*

2.1 INTRODUCTION

Burden assessment plays an increasingly important and accepted role in food safety decision making. Burden assessment is a top–down approach that uses available epidemiological data, for example, generated through surveillance systems, to generate estimates of the health and economic impact of the concerned foodborne disease. These estimates can be used to generate an evidence-based ranking of the impact of foodborne diseases (i.e., risk ranking). Increasingly, these estimates are used to provide justification for the need to strengthen support for increased surveillance and prevention of foodborne diseases by national or international organizations, such as the World Health Organization (WHO). By generating burden estimates at multiple time points, it becomes possible to monitor and evaluate food safety measures over time, as well (Buzby and Roberts, 2009). Finally, health and economic impact may be combined in cost-effectiveness studies that allow to determine the intervention that offers the best value for money invested, so that resources are appropriately allocated (Oostvogels et al., 2015).

In this Chapter we review studies on the health and economic impact of *Campylobacter* at a global and national level.

2.2 HEALTH IMPACT OF *CAMPYLOBACTER*
2.2.1 QUANTIFYING HEALTH IMPACT

Quantifying health impact may be based on disease occurrence (prevalence or incidence), or on the number of deaths (mortality). However, these simple measures of population health do not provide a complete picture of the impact of foodborne

Campylobacter. http://dx.doi.org/10.1016/B978-0-12-803623-5.00002-2

diseases on human health (Mangen et al., 2010; Devleesschauwer et al., 2015). On the one hand, these measures either quantify the impacts of morbidity or mortality, thus prohibiting a comparative ranking of highly morbid but not necessarily fatal diseases and highly lethal diseases. On the other hand, they only quantify occurrence of illness or death, but not severity of illness or death. Indeed, foodborne illnesses may differ in clinical impact and duration of the concerned symptoms. Likewise, ignoring the age at which people die, and thus not considering how many years of healthy life might be lost due to death, does not fairly capture the impact of mortality.

To overcome the limitations of these simple measures, summary measures of population health (SMPHs) have been developed as an additional source of information for measuring disease burden. The disability-adjusted life year (DALY) is currently the most widely used SMPH in public health research. Originally developed to quantify and compare the burden of diseases, injuries, and risk factors within and across countries, the DALY summarizes the occurrence and impact of morbidity and mortality in a single measure (Murray and Lopez, 2013; Devleesschauwer et al., 2014a). The DALY is the key measure in the global burden of disease (GBD) studies, and is officially adopted by the WHO for reporting on health information (Murray et al., 2012; WHO, 2013).

The DALY is a health gap measure, measuring the healthy life years lost due to diseases or injury in a population. DALYs are calculated by adding the number of years of life lost due to premature mortality (YLLs) and the number of years lived with disability, adjusted for severity (YLDs). YLLs are the product of the number of deaths and the residual life expectancy at the age of death. Following an incidence perspective, YLDs are defined as the product of the number of incident cases, the duration until remission or death, and the disability weight that reflects the reduction in health-related quality of life on a scale from zero (full health) to one (death). The incidence perspective assigns all health outcomes, including those in future years, to the initial event (e.g., *Campylobacter* infection). This approach therefore reflects the future burden of disease resulting from current events. An alternative formula for calculating YLDs follows a prevalence perspective, and defines YLDs as the product of the number of prevalent cases with the disability weight (Murray et al., 2012). In this prevalence perspective, the health status of a population is assessed at a specific point in time, and prevalent diseases are attributed to events that happened in the past. This approach thus reflects the current burden of disease resulting from previous events. Although both perspectives are valid, the incidence perspective is more sensitive to current epidemiological trends (Murray, 1994), including the effects of intervention measures, and therefore often preferred for assessment of the burden of foodborne diseases (Devleesschauwer et al., 2015).

Different approaches can be taken for calculating DALYs, depending on whether the interest lies in quantifying the burden of a health outcome, a hazard, or a risk factor (Devleesschauwer et al., 2014b). A natural choice for quantifying the health impact of foodborne diseases is the hazard-based approach. This approach defines the burden of a specific foodborne disease as that resulting from all health states,

FIGURE 2.1 *Campylobacter* DALY Per 100,000 People by Subregion, 2010 (Havelaar et al., 2015)

that is, acute symptoms, chronic sequelae, and death, that are causally related to the concerned hazard, and that may become manifest at different time scales, or have different severity levels (Mangen et al., 2013). The starting point for quantifying DALYs therefore is typically the construction of a disease model or outcome tree that is a schematic representation of the various health states associated with the concerned hazard, and the possible transitions between these states (Devleesschauwer et al., 2014b). As reviewed in Chapter 1, the most important sequelae associated with *Campylobacter* infection are Guillain–Barré syndrome (GBS), reactive arthritis (ReA), inflammatory bowel disease (IBD), and irritable bowel syndrome (IBS).

2.2.2 GLOBAL BURDEN OF *CAMPYLOBACTER*

To date, the most comprehensive assessment of the global burden of campylobacteriosis is the one performed by the Foodborne Disease Burden Epidemiology Reference Group (FERG) of the WHO (Havelaar et al., 2015). FERG estimated that, in 2010, *Campylobacter* was responsible for 166 million [95% Uncertainty Interval (UI) 92–301 million] diarrheal illnesses (of a total of nearly 2 billion attributed diarrheal illnesses), and 31,700 (95% UI 25,400–40,200) GBS cases. These illnesses resulted in 37,600 deaths (95% UI 27,700–55,100), and 3.7 million DALYs [95% UI 2.9–5.2 million; equivalent to 54 (95% UI 42–77) DALYs/100,000] (Kirk et al., 2015). Foodborne transmission was estimated to contribute to 58% (44–69%) of the global disease burden (Hald et al., 2016).

Fig. 2.1 shows the regional variation of the *Campylobacter* burden. The African regions bore nearly half of the global burden, followed by the South-East Asian regions. Of note, while *Campylobacter* was only the sixth most important contributor to the global burden of foodborne disease, it was the most important foodborne hazard in the high-income countries of the American and Western Pacific regions, and the second most important foodborne hazard in the high-income countries of the European region (Havelaar et al., 2015).

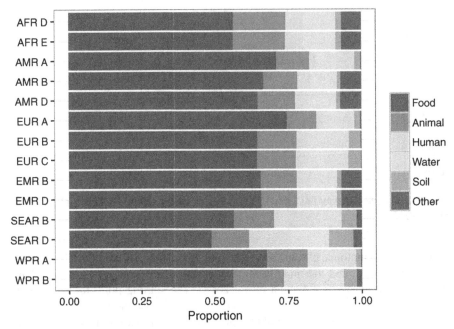

FIGURE 2.2 *Campylobacter* **Exposure Routes by Subregion, 2010 (Hald et al., 2016)**

To estimate the global burden of campylobacteriosis, FERG attributed diarrheal incidence and mortality rates to *Campylobacter* using etiological fractions obtained through metaanalysis (Pires et al., 2015). This information was combined with *Campylobacter* incidence and mortality estimates available for high-income countries. Furthermore, it was estimated that 31% (range 28–45%) of GBS cases globally were associated with antecedent *Campylobacter* infection, and that the GBS case–fatality ratio was 4.1% (range 2.4–6.0%) (Kirk et al., 2015). Other sequelae, such as reactive arthritis, inflammatory bowel disease, and irritable bowel syndrome, were not included in the FERG estimates due to a lack of global data. As national burden studies have shown that sequelae add significantly to the *Campylobacter* disease burden (Mangen et al., 2015), the FERG estimates thus underestimate the true global burden of *Campylobacter*. The contribution of different transmission routes to the *Campylobacter* burden was assessed through a structured expert elicitation study (Hald et al., 2016). Fig. 2.2 shows the resulting regional attribution estimates, highlighting the importance of food as a major transmission route, followed by water, and direct animal contact. Within the foodborne transmission route, poultry was considered to be the dominant source of infection in all regions (Fig. 2.3).

A second source of GBD estimates of *Campylobacter* is the GBD 2013 study conducted by the Institute for Health Metrics and Evaluation. In GBD 2013, only *Campylobacter* enteritis was included. In 2013, *Campylobacter* enteritis was estimated

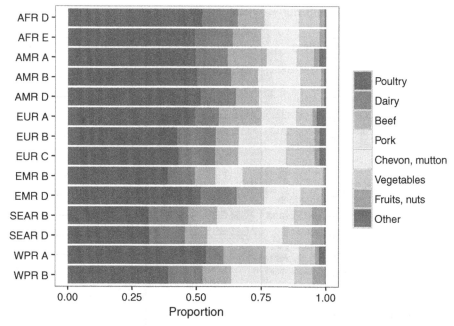

FIGURE 2.3 Sources of Foodborne *Campylobacter* Infection by Subregion, 2010 (Hald et al., 2016)

to be responsible for 14,100 deaths (95% UI 6900–22,400), and 1 million DALYs (95% UI 0.5–1.6 million), or 1.1% (95% UI 0.6–1.8%) of global diarrhea deaths, and 1.4% (95% UI 0.7–2.2%) of global diarrhea DALYs. Of note, the burden of *Campylobacter* enteritis appears to have halved since 1990, where it was estimated to be responsible for 28,400 deaths (95% UI 16,400–42,800), and 2.1 million DALYs (95% UI 1.2–3.2 million).

2.2.3 NATIONAL BURDEN OF *CAMPYLOBACTER*

Since the first national DALY calculation for *Campylobacter* published by Havelaar et al. (2000), several authors have estimated the burden of *Campylobacter* at a national or regional level. Table 2.1 provides an overview of available studies. All studies were performed in high-income countries, and confirmed the importance of *Campylobacter* as a foodborne pathogen. Indeed, when DALY estimates were used to rank multiple foodborne diseases, *Campylobacter* was consistently ranked first or second, with the exception of Greece, where it was ranked seventh (Gkogka et al., 2011). At a population level, the estimated burden of *Campylobacter* ranged from 0 DALYs/100,000 in Cyprus to 82 DALYs/100,000 in Australia. Comparisons across studies should nevertheless be done with caution, given the methodological differences, such as the comprehensiveness of data sources, or the nature of included

Table 2.1 National *Campylobacter* Burden Studies

References	Setting	Included Symptoms and Sequelae	DALY/ 100,000	DALY/ Case
Havelaar et al. (2000)	Netherlands	AGE; GBS; ReA	9.1	0.005
van den Brandhof et al. (2004)	Netherlands	AGE	8.5	0.014
Mangen et al. (2005)	Netherlands	AGE; GBS; ReA; IBD	7.5	0.015
van Lier et al. (2007); Kretzschmar et al. (2012)	Europe (20 countries)	AGE; GBS; ReA; IBD	5.3 (ranging from 0 in Cyprus to 28 in Czech Republic)	0.112
Haagsma et al. (2010)	The Netherlands	IBS	8.6	0.018
Lake et al. (2010)	New Zealand	AGE; GBS; ReA; IBD	4.0	0.013
Ruzante et al. (2010)	Canada, associated with chicken consumption	AGE; GBS	2.3	0.005
Gkogka et al. (2011)	Greece	AGE; GBS; ReA; IBD; IBS	0.5	0.001
Havelaar et al. (2012)	The Netherlands	AGE; GBS; ReA; IBD; IBS	20	0.041
Hoffmann et al. (2012); Batz et al. (2012, 2014)	USA	AGE; GBS	4.6[a]	0.016[a]
Toljander et al. (2012)	Sweden	AGE; GBS; ReA	2.8	0.004
Werber et al. (2013)	Germany	Only years of life lost were estimated	0.4	0.001
Kwong et al. (2012)	Canada		0.5	0.002
Gibney et al. (2014)	Australia	AGE; GBS; ReA; IBS	82	0.024
Kumagai et al. (2015)	Japan	AGE; GBS; ReA; IBD	4.8	0.051
Mangen et al. (2015)	The Netherlands	AGE; GBS; ReA; IBD; IBS	22	0.039
Scallan et al. (2015)	USA	AGE; GBS; ReA; IBS	7.5	0.027

[a]*Estimates are QALY losses instead of DALYs*

sequelae. Indeed, nearly all studies included GBS as a sequela, whereas the inclusion of ReA, IBD, and IBS was more variable. Haagsma et al. (2010) specifically looked at the burden of postinfectious IBS in The Netherlands, estimated to occur in 9% of *Campylobacter* patients, and found that including IBS doubled the burden estimate for *Campylobacter*. Across studies, the estimated DALY/case ranged from 0.001 to 0.112, corresponding to an average loss of less than 1 to 41 days of healthy life. However, for individual patients with specific sequelae, the health loss can be much more significant. Residual symptoms of GBS are, for instance, estimated to result in a loss of more than 6 years of healthy life (Havelaar et al., 2000).

2.3 ECONOMIC IMPACT OF *CAMPYLOBACTER*
2.3.1 COST-OF-ILLNESS

Foodborne diseases incur costs associated with illness and death, and impose an economic burden to the food industry, and the regulatory and public health sectors (Buzby and Roberts, 2009). Historically, however, most economic impact assessments only focused on the societal cost of human illness and death (Mangen et al., 2015). In these cost-of-illness (COI) studies, a distinction is typically made between direct and indirect costs, on the one hand, and healthcare and nonhealthcare costs, on the other (Mangen et al., 2010). Direct healthcare costs relate to the resources provided by the healthcare sector, such as healthcare provider consultations, diagnosis, medication, and hospitalization. Direct nonhealthcare costs (also called patient costs) relate to the resources used for healthcare borne by the patient and/or his family, such as over-the-counter medications, patient copayments for healthcare, and travel expenses to visit a healthcare provider. Indirect nonhealthcare costs mostly include productivity losses due to absenteeism, or job loss of patients and their caregivers. Indirect healthcare costs relate to medical consumption in life-years gained due to life-saving or death-postponing interventions, and are therefore by definition not included in COI studies.

COI estimates for *Campylobacter* have been generated since the 1980s (Todd, 1989), yet no estimate of the global economic impact of *Campylobacter* is available. Table 2.2 gives an overview of available *Campylobacter* COI estimates. All studies were performed in developed countries. Across countries, the indirect nonhealthcare costs (i.e., mostly productivity losses) appear to be the greatest contributor to the overall *Campylobacter* COI (Buzby et al., 1997; Roberts et al., 2003; Lake et al., 2010; Mangen et al., 2015). As for health impact assessment studies, it is important to include chronic sequelae in COI studies, as they can result in high individual and total costs (Buzby and Roberts, 2009; Mangen et al., 2015). Buzby et al. (1997) appear to be to first to evaluate COI of *Campylobacter*-associated GBS. They estimated that in 1995, in the USA, *Campylobacter* resulted in a total annual cost of US $1.5–8.0 billion, of which US $0.2–1.8 billion was due to *Campylobacter*-associated GBS (or around US $470,000/patient). The productivity cost of GBS patients not able to resume work was the largest contributor to the overall COI of *Campylobacter*-associated GBS.

However, comparison between studies is difficult. There exists, for example, no universally accepted method to estimate productivity losses in case of illness-related death or permanent disability (Buzby et al., 1997). The more frequently used methodology for estimating productivity losses due to absence from paid and unpaid work is the human capital approach, based on neoclassical labor theory. In the human capital approach, the value of potential lost income because of illness-related death or permanent disability is estimated, starting from the age of death or permanent disability, up to the age of retirement. However, arguing that the neoclassical labor theory is out of line with reality to current labor markets, Koopmanschap et al. (1995) introduced the friction cost approach. In this approach, productivity losses are only

Table 2.2 National *Campylobacter* Cost-of-Illness (COI) Studies

References	Setting	Reference Year	Included Symptoms and Sequelae	Included Cost Components	Total COI	COI/Case
Todd (1989), Roberts (1986)	United States	Not indicated	AGE	DHC; PL$_H$ and intangible cost (death)	Up to US $1.4 billion	Up to US $666
Buzby et al. (1997)	USA	1995	AGE; GBS	DHC; PL$_H$ and intangible costs (death/disability)	US $1.5–8.0 billion	US $750–800
Withington and Chambers (1997)	New Zealand	1995	AGE; GBS; ReA	DHC; DNHC; PL$_H$	NZ $4.5 million	NZ $596
Scott et al. (2000)	New Zealand	1999	AGE	DHC; DNHC; PL$_H$ and intangible costs (death)	NZ $40 million	NZ $533
Roberts et al. (2003)	United Kingdom	1993–1995	AGE	DHC; DNHC; PL$_H$	£70 million	£315
van den Brandhof et al. (2004)	Netherlands	1999	AGE	DHC; DNHC; PL$_F$	€9.2 million	€103
Mangen et al. (2005)	The Netherlands	2000	AGE; GBS; ReA; IBS	DHC; DNHC; PL$_F$	€21 million	€233
Gellynck et al. (2008)	Belgium	2004	AGE; GBS; ReA; IBD	DHC; DNHC; PL$_F$	€27 million	€495
Scharff et al. (2009)	Ohio, USA		AGE; GBS	DHC; PL$_H$ and intangible cost (morbidity and death)	US $217 million	US $3411
Lake et al. (2010)	New Zealand	2006	AGE; GBS; ReA; IBD	DHC; DNHC; PL$_H$	NZ $134 million	NZ $600
Ruzante et al. (2010)	Canada, associated with chicken consumption	2006	AGE; GBS	DHC; PL$_H$ and intangible costs (death)	CAN $80 million	CAN $512
Collier et al. (2012)	USA	2007	AGE (only hospitalized cases)	DHC (including copayments by patients)	US $118 million	US $8915
Scharff (2012)	USA	2010	AGE; GBS	DHC; PL$_H$ and intangible cost (morbidity and death)	US $1.56 billion	US $1846
Mangen et al. (2015)	Netherlands	2011	AGE; GBS; ReA; IBD; IBS	DHC; DNHC; PL$_F$	€82 million	€757
Tam and O'Brien (2016)	United Kingdom	2008–2009	AGE; GBS	DHC; DNHC	£51 million	£90

Abbreviations: AGE, acute gastro-enteritis; GBS, Guillain–Barré syndrome; IBD, inflammatory bowel disease; IBS, irritable bowel syndrome; ReA, reactive arthritis; DHC, direct healthcare costs; DNHC, direct nonhealthcare costs, also called patient costs; PL$_F$, productivity losses using the friction cost approach; PL$_H$, productivity losses using the human capital approach;
When monetarizing intangible costs, most studies used national values for a statistical life (Todd, 1989; Buzby et al., 1997; Scott et al., 2000; Scharff et al., 2009; Scharff, 2012; Lake et al., 2010; Ruzante et al., 2010). Scharff et al. (2009) further made an estimate for intangible costs by monetarizing QALYs
Scharff (2012); Lake et al., 2010; Ruzante et al., 2010).

considered for the period needed to replace a sick, invalid, or deceased worker, the so-called friction period that depends on the situation on the labor market, and places a zero value on individuals outside of the labor market, that is, children, retirees, and the elderly (Koopmanschap and van Ineveld, 1992).

A further complicating fact when comparing cost studies is that some studies considered both the financial impact of the disease (e.g., medical costs, patient expenses and productivity losses), and intangible costs for suffering, bad health, and premature death. Intangible costs are monetarized by using revealed or stated preferences of willingness-to-pay (WTP) (Drummond et al., 2015). WTP measures what individuals would be willing to pay to obtain health improvements, or to avoid adverse health states (Krupnick, 2004; Drummond et al., 2015). WTP can be measured by evaluating the trade-offs people actually make (revealed preferences), or by presenting people with hypothetical choices (stated preferences) (Krupnick, 2004; Drummond et al., 2015). This method is based on the trade-offs that individuals must make between health and other goods and is consistent, therefore, with the theoretical foundation of welfare economics (Drummond et al., 2015). Such trade-offs between money and fatality risks serve to estimate the value of a statistical life (Viscusi and Aldy, 2003).

Another attempt to monetarize intangible costs such as bad health and premature death was done by Scharff et al. (2009), who proposed an enhanced COI model that incorporated a value for pain and suffering. This value was calculated by monetizing losses in quality-adjusted life years (QALYs). QALY losses are roughly similar to DALYs, and are based in part on functional disability, pain, and suffering. The monetization of QALY losses was based on the assumption that one QALY is worth the value of a statistical life year. When applied to the entire USA, the enhanced COI model resulted in an estimated *Campylobacter* COI of US $8141/case, or US $6.9 billion in total, significantly higher than the estimates of the basic COI model that only considered financial impact (US $1846/case, or US $1.56 billion in total) (Scharff, 2012).

2.3.2 INDUSTRY AND GOVERNMENT COSTS

Even though COI appears to be the dominant approach for estimating the economic impact of foodborne diseases, there are various other economic losses beyond those resulting from human illness (Buzby and Roberts, 2009). Indeed, surveillance and other regulatory activities in place to monitor, prevent, and control foodborne diseases incur cost to the society.

Incidental foodborne disease outbreaks are not just associated with a peak in human illnesses, and thus a peak in COIs, they result in additional economic consequences due to costs of investigation, law suits, and loss of business by the food company (e.g., due to recalls, loss of consumer trust, or trade restrictions) (Todd, 1989). Sheerin et al. (2014), for example, estimated that a waterborne outbreak of campylobacteriosis in Darfield, New Zealand, imposed an additional NZ $95,000 to the District Council, due to additional staff time, and a commissioned investigation report.

To tackle the New Zealand *Campylobacter* epidemic, new *Campylobacter* compliance standards were imposed to the industry in 2007. Industry costs of capital investment were estimated at NZ $2 million, while increased operating costs, including purchase of chemicals and maintenance costs, were determined to be NZ $0.88 million. The new compliance program required the regulator to undertake and continue oversight of its implementation, imposing an additional annual cost on the government of NZ $0.89 million (Duncan, 2014).

Further indications on potential industry and government costs are available from cost-effectiveness studies (Havelaar et al., 2007; Elliott et al., 2012; Lake et al., 2013). The CARMA (Campylobacter Risk Management and Assessment) project aimed to assess the cost–utility of different potential *Campylobacter* intervention measures in The Netherlands. Estimates were generated of the presumed direct intervention costs and number of *Campylobacter* gastroenteritis cases averted, allowing for the calculation of cost–utility ratios (Havelaar et al., 2007). Thus, the cost of improved farm hygiene was estimated at €8–63 million, and the costs of information campaigns, for example, to stimulate hygienic kitchen behavior, or to promote home freezing of poultry, were estimated at €1 million/year. Scheduled decontamination of carcasses by dipping in lactic acid would cost €5 million, and avert 9200 *Campylobacter* cases, resulting in a cost of €28,000/DALY averted—which was found to be the most beneficial cost–utility ratio. Irradiation, probably the most effective intervention, was considered to be too expensive and therefore not cost-effective. Not considered in these estimates were potential indirect effects due to considered interventions, for example, the nonacceptance by the consumer, the loss of market shares as not being able to sell on time. The consequence would be lower selling prices (i.e., lower income for the industry) and consequently less cost-effective interventions.

Finally, for zoonotic foodborne diseases, livestock production losses due to clinical or subclinical infection may further add to the economic burden. Poultry infected with *Campylobacter*, however, are generally neither sick, nor are their growth and reproduction abilities affected (Mangen et al., 2007). Implementing farm-level interventions to control *Campylobacter* would thus result in a net rise of production costs equal to the direct intervention cost. This skewed situation might impede program uptake, and would call for governments to intervene, as the guardian of food safety.

2.4 CONCLUDING REMARKS

Several studies have focused on producing DALY and cost estimates associated with foodborne *Campylobacter* infections. From a methodological standpoint, approaches and data characteristics used for the estimates differed in both fields, making direct comparisons difficult. Alignment of approaches and methodologies would be an important future step. Nevertheless, the importance of *Campylobacter* in the foodborne disease burden was consistently shown in these studies, warranting continuing efforts the reduce food contamination by this pathogen.

REFERENCES

Batz, M.B., Hoffmann, S., Morris, Jr., J.G., 2012. Ranking the disease burden of 14 pathogens in food sources in the United States using attribution data from outbreak investigations and expert elicitation. J. Food Prot. 75 (7), 1278–1291.

Batz, M., Hoffmann, S., Morris, Jr., J.G., 2014. Disease-outcome trees, EQ-5D scores, and estimated annual losses of quality-adjusted life years (QALYs) for 14 foodborne pathogens in the United States. Foodborne Pathog. Dis. 11 (5), 395–402.

Buzby, J.C., Roberts, T., 2009. The economics of enteric infections: human foodborne disease costs. Gastroenterology 136 (6), 1851–1862.

Buzby, J.C., Allos, B.M., Roberts, T., 1997. The economic burden of *Campylobacter*-associated Guillain-Barré syndrome. J. Infect. Dis. 176 (Suppl. 2), S192–S197.

Collier, S.A., Stockman, L.J., Hicks, L.A., Garrison, L.E., Zhou, F.J., Beach, M.J., 2012. Direct healthcare costs of selected diseases primarily or partially transmitted by water. Epidemiol. Infect. 140 (11), 2003–2013.

Devleesschauwer, B., Havelaar, A.H., Maertens de Noordhout, C., Haagsma, J.A., Praet, N., Dorny, P., Duchateau, L., Torgerson, P.R., Van Oyen, H., Speybroeck, N., 2014a. Calculating disability-adjusted life years to quantify burden of disease. Int. J. Public Health 59 (3), 565–569.

Devleesschauwer, B., Havelaar, A.H., Maertens de Noordhout, C., Haagsma, J.A., Praet, N., Dorny, P., Duchateau, L., Torgerson, P.R., Van Oyen, H., Speybroeck, N., 2014b. DALY calculation in practice: a stepwise approach. Int. J. Public Health 59 (3), 571–574.

Devleesschauwer, B., Haagsma, J.A., Angulo, F.J., Bellinger, D.C., Cole, D., Döpfer, D., Fazil, A., Fèvre, E.M., Gibb, H.J., Hald, T., Kirk, M.D., Lake, R.J., Maertens de Noordhout, C., Mathers, C.D., McDonald, S.A., Pires, S.M., Speybroeck, N., Thomas, M.K., Torgerson, P.R., Wu, F., Havelaar, A.H., Praet, N., 2015. Methodological framework for World Health Organization estimates of the global burden of foodborne disease. PLoS One 10 (12), e0142498.

Drummond, M.F., Sculpher, M.J., Claxton, K., Stoddart, G.L., Torrance, G.W., 2015. Methods for the economic evaluation of health care programmes, fourth ed. Oxford University Press, Oxford.

Duncan, G.E., 2014. Determining the health benefits of poultry industry compliance measures: the case of campylobacteriosis regulation in New Zealand. N.Z. Med. J. 127 (1391), 22–37.

Elliott, J., Lee, D., Erbilgic, A., Jarvis, A., 2012. Analysis of the costs and benefits of setting certain control measures for reduction of *Campylobacter* in broiler meat at different stages of the food chain. Report submitted by ICF GHK in association with ADAS. Available from: http://ec.europa.eu/food/food/biosafety/salmonella/docs/campylobacter_cost_benefit_analysis_en.pdf

Gellynck, X., Messens, W., Halet, D., Grijspeerdt, K., Hartnett, E., Viaene, J., 2008. Economics of reducing *Campylobacter* at different levels within the Belgian poultry meat chain. J. Food Prot. 71 (3), 479–485.

Gibney, K.B., O'Toole, J., Sinclair, M., Leder, K., 2014. Disease burden of selected gastrointestinal pathogens in Australia, 2010. Int. J. Infect. Dis. 28, 176–185.

Gkogka, E., Reij, M.W., Havelaar, A.H., Zwietering, M.H., Gorris, L.G., 2011. Risk-based estimate of effect of foodborne diseases on public health, Greece. Emerg. Infect. Dis. 17 (9), 1581–1590.

Haagsma, J.A., Siersema, P.D., De Wit, N.J., Havelaar, A.H., 2010. Disease burden of postinfectious irritable bowel syndrome in The Netherlands. Epidemiol. Infect. 138 (11), 1650–1656.

Hald, T., Aspinall, W., Devleesschauwer, B., Cooke, R., Corrigan, T., Havelaar, A.H., Gibb, H.J., Torgerson, P.R., Kirk, M.D., Angulo, F.J., Lake, R.J., Speybroeck, N., Hoffmann, S., 2016. World Health Organization estimates of the relative contributions of food to the burden of disease due to selected foodborne hazards: a structured expert elicitation. PLoS One 11 (1), e0145839.

Havelaar, A.H., de Wit, M.A., van Koningsveld, R., van Kempen, E., 2000. Health burden in the Netherlands due to infection with thermophilic *Campylobacter* spp. Epidemiol. Infect. 125 (3), 505–522.

Havelaar, A.H., Mangen, M.J., de Koeijer, A.A., Bogaardt, M.J., Evers, E.G., Jacobs-Reitsma, W.F., van Pelt, W., Wagenaar, J.A., de Wit, G.A., van der Zee, H., Nauta, M.J., 2007. Effectiveness and efficiency of controlling *Campylobacter* on broiler chicken meat. Risk Anal. 27 (4), 831–844.

Havelaar, A.H., Haagsma, J.A., Mangen, M.J., Kemmeren, J.M., Verhoef, L.P., Vijgen, S.M., Wilson, M., Friesema, I.H., Kortbeek, L.M., van Duynhoven, Y.T., van Pelt, W., 2012. Disease burden of foodborne pathogens in the Netherlands, 2009. Int. J. Food Microbiol. 156 (3), 231–238.

Havelaar, A.H., Kirk, M.D., Torgerson, P.R., Gibb, H.J., Hald, T., Lake, R.J., Praet, N., Bellinger, D.C., de Silva, N.R., Gargouri, N., Speybroeck, N., Cawthorne, A., Mathers, C., Stein, C., Angulo, F.J., Devleesschauwer, B., World Health Organization Foodborne Disease Burden Epidemiology Reference Group, 2015. World Health Organization global estimates and regional comparisons of the burden of foodborne disease in 2010. PLoS Med. 12 (12), e1001923.

Hoffmann, S., Batz, M.B., Morris, Jr., J.G., 2012. Annual cost of illness and quality-adjusted life year losses in the United States due to 14 foodborne pathogens. J. Food Prot. 75 (7), 1292–1302.

Kirk, M.D., Pires, S.M., Black, R.E., Caipo, M., Crump, J.A., Devleesschauwer, B., Döpfer, D., Fazil, A., Fischer-Walker, C.L., Hald, T., Hall, A.J., Keddy, K.H., Lake, R.J., Lanata, C.F., Torgerson, P.R., Havelaar, A.H., Angulo, F.J., 2015. World Health Organization estimates of the global and regional disease burden of 22 foodborne bacterial, protozoal, and viral diseases, 2010: a data synthesis. PLoS Med. 12 (12), e1001921.

Koopmanschap, M.A., van Ineveld, B.M., 1992. Towards a new approach for estimating indirect costs of disease. Soc. Sci. Med. 34 (9), 1005–1010.

Koopmanschap, M.A., Rutten, F.F.H., Van Ineveld, B.M., Van Roijen, L., 1995. The friction cost method for measuring indirect costs of disease. J. Health Econ. 14 (2), 171–189.

Kretzschmar, M., Mangen, M.J., Pinheiro, P., Jahn, B., Fèvre, E.M., Longhi, S., Lai, T., Havelaar, A.H., Stein, C., Cassini, A., Kramarz, P., BCoDE consortium, 2012. New methodology for estimating the burden of infectious diseases in Europe. PLoS Med. 9 (4), e1001205.

Krupnick, A.J., 2004. Valuing health outcomes: policy choices and technical issues. RFF Report. Resources for the Future, Washington, DC.

Kumagai, Y., Gilmour, S., Ota, E., Momose, Y., Onishi, T., Bilano, V.L., Kasuga, F., Sekizaki, T., Shibuya, K., 2015. Estimating the burden of foodborne diseases in Japan. Bull. World Health Organ. 93 (8), 540–549.

Kwong, J.C., Ratnasingham, S., Campitelli, M.A., Daneman, N., Deeks, S.L., Manuel, D.G., Allen, V.G., Bayoumi, A.M., Fazil, A., Fisman, D.N., Gershon, A.S., Gournis, E., Heathcote, E.J., Jamieson, F.B., Jha, P., Khan, K.M., Majowicz, S.E., Mazzulli, T., McGeer, A.J., Muller, M.P., Raut, A., Rea, E., Remis, R.S., Shahin, R., Wright, A.J., Zagorski, B., Crowcroft, N.S., 2012. The impact of infection on population health: results of the Ontario burden of infectious diseases study. PLoS One 7 (9), e44103.

Lake, R.J., Cressey, P.J., Campbell, D.M., Oakley, E., 2010. Risk ranking for foodborne microbial hazards in New Zealand: burden of disease estimates. Risk Anal. 30 (5), 743–752.

Lake, R.J., Horn, B.J., Dunn, A.H., Parris, R., Green, F.T., McNickle, D.C., 2013. Cost-effectiveness of interventions to control *Campylobacter* in the New Zealand poultry meat food supply. J. Food Prot. 76 (7), 1161–1167.

Mangen, M.J.J., Havelaar, A.H., Bernsen, R.A.J.A.M., Van Koningsveld, R., De Wit, G.A., 2005. The costs of human *Campylobacter* infections and sequelae in the Netherlands: A DALY and cost-of-illness approach. Food Econ. Acta Agr. Scand. Sect. C 2 (1), 35–51.

Mangen, M.J.J., de Wit, G.A., Havelaar, A.H., 2007. Economic analysis of *Campylobacter* control in the Dutch broiler meat chain. Agribusiness 23 (2), 173–192.

Mangen, M.J., Batz, M.B., Käsbohrer, A., Hald, T., Morris, J.G., Taylor, M., Havelaar, A.H., 2010. Integrated approaches for the public health prioritization of foodborne and zoonotic pathogens. Risk Anal. 30 (5), 782–797.

Mangen, M.J., Plass, D., Havelaar, A.H., Gibbons, C.L., Cassini, A., Mühlberger, N., van Lier, A., Haagsma, J.A., Brooke, R.J., Lai, T., de Waure, C., Kramarz, P., Kretzschmar, M.E., BCoDE Consortium, 2013. The pathogen- and incidence-based DALY approach: an appropriate [corrected] methodology for estimating the burden of infectious diseases. PLoS One 8 (11), e79740.

Mangen, M.J., Bouwknegt, M., Friesema, I.H., Haagsma, J.A., Kortbeek, L.M., Tariq, L., Wilson, M., van Pelt, W., Havelaar, A.H., 2015. Cost-of-illness and disease burden of food-related pathogens in the Netherlands, 2011. Int. J. Food Microbiol. 196, 84–93.

Murray, C.J., 1994. Quantifying the burden of disease: the technical basis for disability-adjusted life years. Bull. World Health Organ. 72 (3), 429–445.

Murray, C.J., Lopez, A.D., 2013. Measuring the global burden of disease. N. Engl. J. Med. 369 (5), 448–457.

Murray, C.J., Ezzati, M., Flaxman, A.D., Lim, S., Lozano, R., Michaud, C., Naghavi, M., Salomon, J.A., Shibuya, K., Vos, T., Wikler, D., Lopez, A.D., 2012. GBD 2010: design, definitions, and metrics. Lancet 380 (9859), 2063–2066.

Oostvogels, A.J., De Wit, G.A., Jahn, B., Cassini, A., Colzani, E., De Waure, C., Kretzschmar, M.E., Siebert, U., Mühlberger, N., Mangen, M.J., 2015. Use of DALYs in economic analyses on interventions for infectious diseases: a systematic review. Epidemiol. Infect. 143 (9), 1791–1802.

Pires, S.M., Fischer-Walker, C.L., Lanata, C.F., Devleesschauwer, B., Hall, A.J., Kirk, M.D., Duarte, A.S., Black, R.E., Angulo, F.J., 2015. Aetiology-specific estimates of the global and regional incidence and mortality of diarrhoeal diseases commonly transmitted through food. PLoS One 10 (12), e0142927.

Roberts, T., 1986. The economic losses due to selected foodborne diseases. In: Proceedings of the Ninetieth Annual Meeting of the United States Animal Health Association. pp. 336–353.

Roberts, J.A., Cumberland, P., Sockett, P.N., Wheeler, J., Rodrigues, L.C., Sethi, D., Roderick, P.J., Infectious Intestinal Disease Study Executive, 2003. The study of infectious intestinal disease in England: socio-economic impact. Epidemiol. Infect. 130 (1), 1–11.

Ruzante, J.M., Davidson, V.J., Caswell, J., Fazil, A., Cranfield, J.A., Henson, S.J., Anders, S.M., Schmidt, C., Farber, J.M., 2010. A multifactorial risk prioritization framework for foodborne pathogens. Risk Anal. 30 (5), 724–742.

Scallan, E., Hoekstra, R.M., Mahon, B.E., Jones, T.F., Griffin, P.M., 2015. An assessment of the human health impact of seven leading foodborne pathogens in the United States using disability adjusted life years. Epidemiol. Infect. 143 (13), 2795–2804.

Scharff, R.L., 2012. Economic burden from health losses due to foodborne illness in the United States. J. Food Prot. 75 (1), 123–131.

Scharff, R.L., McDowell, J., Medeiros, L., 2009. Economic cost of foodborne illness in Ohio. J. Food Prot. 72 (1), 128–136.

Scott, W.G., Scott, H.M., Lake, R.J., Baker, M.G., 2000. Economic cost to New Zealand of foodborne infectious disease. N.Z. Med. J. 113 (1113), 281–284.

Sheerin, I., Bartholomew, N., Brunton, C., 2014. Estimated community costs of an outbreak of campylobacteriosis resulting from contamination of a public water supply in Darfield, New Zealand. N.Z. Med. J. 127 (1391), 13–21.

Tam, C.C., O'Brien, S.J., 2016. Economic cost of *Campylobacter*, Norovirus and Rotavirus disease in the United Kingdom. PLoS One 11 (2), e0138526.

Todd, E.C., 1989. Costs of acute bacterial foodborne disease in Canada and the United States. Int. J. Food Microbiol. 9 (4), 313–326.

Toljander, J., Dovärn, A., Andersson, Y., Ivarsson, S., Lindqvist, R., 2012. Public health burden due to infections by verocytotoxin-producing *Escherichia coli* (VTEC) and *Campylobacter* spp. as estimated by cost of illness and different approaches to model disability-adjusted life years. Scand. J. Public Health 40 (3), 294–302.

van den Brandhof, W.E., De Wit, G.A., de Wit, M.A., van Duynhoven, Y.T., 2004. Costs of gastroenteritis in The Netherlands. Epidemiol. Infect. 132 (2), 211–221.

van Lier, E.A., Havelaar, A.H., Nanda, A., 2007. The burden of infectious diseases in Europe: a pilot study. Euro Surveill. 12 (12), E3–E4.

Viscusi, W.K., Aldy, J.E., 2003. The value of a statistical life: a critical review of market estimates throughout the world. J. Risk Uncertainty 27 (1), 5–76.

Werber, D., Hille, K., Frank, C., Dehnert, M., Altmann, D., Müller-Nordhorn, J., Koch, J., Stark, K., 2013. Years of potential life lost for six major enteric pathogens, Germany, 2004–2008. Epidemiol. Infect. 141 (5), 961–968.

World Health Organization, 2013. WHO methods and data sources for global burden of disease estimates 2000–2011. Global health estimates technical paper. WHO/HIS/HSI/GHE/2013.4. Available from: http://www.who.int/healthinfo/statistics/GlobalDALYmethods_2000_2011.pdf

Withington, S.G., Chambers, S.T., 1997. The cost of campylobacteriosis in New Zealand in 1995. N.Z. Med. J. 110, 222–224.

Taxonomy and physiological characteristics of *Campylobacter* spp.

3

Sati Samuel Ngulukun

Bacterial Research Department, National Veterinary Research Institute,
Vom, Plateau State, Nigeria

3.1 INTRODUCTION

The genus *Campylobacter* has expanded considerably since it was initially proposed by Sebald and Veron (1963). In recent years, major progress has been made in understanding the taxonomy of Campylobacters (On and Harrington, 2001; Debruyne et al., 2008). Since then, important new contributions have been published in the area of 16S rRNA and *hsp*60 gene sequence analysis, matrix-assisted laser desorption/ionization time of flight (MALDI-TOF) profile and whole-genome sequence analysis of *Campylobacter* strains to explain to which extent strains within species and strains representing different species differ, and this has led to the description of novel *Campylobacter* species (On, 2013; Koziel et al., 2014; Gilbert et al., 2015).

The aim of this Chapter is to highlight the major developments in the taxonomy of *Campylobacter*, from its inception to the present day. A better understanding of the physiological characteristics of Campylobacters, together with improved molecular techniques, will further elucidate *Campylobacter* biodiversity, and help to isolate and describe new species to expand the taxa further.

3.2 TAXONOMIC HISTORY OF *CAMPYLOBACTER*

McFadyean and Stockman (1913) recorded the first isolation of a *Vibrio*-like organism from aborted fetuses, and named it "*Vibrio fetus*." In 1919, *V. fetus* was also isolated from aborted bovine fetus (Smith and Taylor, 1919). A similar vibrio strain was also recovered from the blood cultures of women who aborted, in 1947 (Vinzent et al., 1947). Jones et al. (1931) and Doyle (1948) isolated a *Vibrio* from the feces of cattle and pigs with diarrhea, respectively. Levi (1946) and King (1957) reported similar *Vibrio* from the blood cultures of humans with gastroenteritis, and Sebald and

Veron (1963) proposed the genus, *Campylobacter*. They transferred these *Vibrio* species, *V. fetus* and *V. bubulus* into the new genus, *Campylobacter*, as *Campylobacter fetus* sp. nov., comb. nov., and *C. bubulus* sp. nov., comb. nov., respectively.

Veron and Chatelain (1973) published a more comprehensive study of the taxonomy of the *Vibro*-like organisms, and proposed four distinctive species in the genus *Campylobacter*: *C. fetus* (the type species), *C. coli* (isolated from pigs), *C. jejuni* (isolated from cattle, human, sheep), and *C. sputorum* with two subspecies, *C. sputorum* subsp. *sputorum* (isolated from the sputum of a patient with bronchitis – Prèvot, 1940), and *Campylobacter sputorum* subsp. *bubulus* (isolated from bovine vagina and semen – Florent, 1953).

During the 1970s, renewed interest in *Campylobacter* followed the recognition of *C. jejuni* and *C. coli* as a cause of diarrhea in humans (Dekeyser et al., 1972). Butzler et al. (1973) applied the filtration techniques used in veterinary microbiology to selectively isolate *Campylobacter* from the stools of humans with diarrhea. A few years later, Skirrow (1977) described a selective supplement to isolate Campylobacters, and evaluate their clinical role.

Improvement in isolation procedures and clinical importance led to renewed interest in *Campylobacter* research during the 1980s. As a result, many *Campylobacter*-like organisms (CLO) were isolated from a variety of sources – human and animal – and the environments and new species were described: *C. concisus* (Tanner et al., 1981), *C. mucosalis* (Lawson et al., 1981; Roop et al., 1985), *C. lari* (Benjamin et al., 1983), *Wolinella recta* (Tanner et al., 1981) and *W. curva* (Tanner et al., 1984), later reclassified as *C. rectus* and *C. curvus* respectively, *C. hyointestinalis* (Gebhart et al., 1985), *C. pyloridis* (Marshall et al., 1984) that was later moved to the genus *Helicobacter* (Goodwin et al., 1989), *C. nitrofigilis* (McLung et al., 1983) that was later moved to the genus *Arcobacter* (Vandamme et al., 1991), *C. cryaerophila* (Neill et al., 1985), also later moved to the genus *Arcobacter*, *C. mustalae* (Fox et al., 1989) later moved to the genus *Helicobacter* (Goodwin et al., 1989), as well as *C. cinaedi* and *C. fenneliae* (Totten et al., 1985), later moved to the genus *Helicobacter*.

Also in the 1980s, the study of bacterial phylogeny using DNA–RNA based technologies became more popular, and was increasingly used to evaluate and revise bacterial classification schemes (Debruyne et al., 2008). In 1985, Roop et al. (1985) first demonstrated, using DNA homology studies, that catalase-negative Campylobacters represent a distinct species. The first RNA-based phylogenetic study of *Campylobacter* species was published by Thompson et al. (1988), in which the phylogenetic relationship of all species in the genus *Campylobacter*, *W. succinogenes*, and other Gram-negative bacteria were determined by comparison of partial 16S rRNA sequences. The result of the study grouped these species into three separate ribosomal RNA sequence homology groups: Homology group I contained the following: *C. fetus* (type species), *C. coli*, *C. jejuni*, *C. lari*, *C. hyointestinalis*, *C. concisus*, *C. mucosalis*, *C. sputorum*, and *C. upsaliensis*. Group II: *C. cinaedi*, *C. fenelliae*, *C. pylori*, and *W. succinogenes*. Group III: *C. cryaerophila* and *C. nitrofigilis*. Goodwin et al. (1989) transferred the gastric species, *C. pylori* and *C. mustelae*, from the genus *Campylobacter* and *Wolinella*, and proposed a novel genus, *Helicobacter*, and named them *H. pylori* and *H. mustelae*, respectively. In 1991, a complete revision of the taxonomy and nomenclature of the genus *Campylobacter* and related bacteria was prepared by Vandamme et al. (1991). They examined in a DNA–rRNA hybridization study over 70 *Campylobacter* strains and related taxa, and performed an immunotyping analysis of 130 antigens versus 34 antisera of *Campylobacter* and related taxa, and found that all of the named *Campylobacter* and related taxa belong to the same taxonomic group, which they named rRNA superfamily VI, and that is far removed from the Gram-negative bacteria allocated to the former rRNA superfamily sensu De Ley (Vandamme and De Ley, 1991). This phylogenetic lineage is now known as the Epsilon subdivision of the proteobacteria. They proposed that the emended *Campylobacter* genus should be limited to *C. fetus*, *C. hyolintestinalis*, *C. concisus*, *C. mucosalis*, *C. sputorum*, *C. jejuni*, *C. coli*, *C. lari*, and *C. upsaliensis*. *W. curva* and *W. recta* were transferred to the genus *Campylobacter* as *C. curvus* and *C. rectus*, respectively. They also observed that *Bacteroides gracilis* and *B. ureolyticus* are generically misnamed, and are closely related to the genus *C. nitrofigilis* and *C. cryaerophila*, and other aerotolerant *Campylobacter* organism (CLO) that were later classified as *Arcobacter butzleri* and *A. skirrowii* (Vandamme et al., 1992). They described a new genus, *Arcobacter*. *W. succinogenes* remained the only species of the genus *Wolinella*. The genus *Helicobacter* was also emended; *C. cinaedi* and *C. finnaelliae* (Totten *et al.,* 1985) were included in this genus as *H. cinaedi* and *H. finnaelliae* respectively. *Flexispira rappini* (Bryner et al., 1986), as the only species, was closely related to the genus *Helicobacter*. *Campylobacter*-like species reclassified later as *Sulfurospirillum* species (Stolz et al., 1999) did not belong to any of these genera, and the authors suggested that they probably constitute a distinct genus within the rRNA superfamily VI. The generically misnamed *B. gracilis* was included in the emended genus, *Campylobacter,* as *C. gracilis* (Vandamme et al., 1995). *B. ureolyticus* was a close relative of the emended genus *Campylobacter*, but its taxonomic status did not change until 2010 (Vandamme et al., 2010), when it was transferred to *C. ureolyticus*. Stanley et al. (1992) isolated and characterized thermophilic catalase-negative *Campylobacter* strains, and named them *C. helveticus*. Etoh et al. (1993, 1998) isolated and characterized *Campylobacter*-like strains from human gingival crevices, and found they were biochemically similar to *C. curvus* and *C. rectus*, but differed from them based on whole cell protein profile and DNA hybridization, and named them C. *showae*. By 1998, On et al. (1998) revised the *C. sputorum* taxon (Veron and Chatelain, 1973), and stated that *C. sputorum* comprises three biovars, defined on the basis of a given strain to produce catalase or urease: *C. sputorum* bv. *sputorum* (negative in both tests), *C. sputorum* bv. *fecalis* (catalase positive, urease negative), and *C. sputorum* bv. *paraureolyticus* (urease positive); *C. sputorum* bv. *bubulus* (Veron and Chatelain, 1973) was reclassified as bv. *sputorum*.

Between 2000 and 2009, several new species were described and added to the taxa. *C. lanienae* was isolated from pigs (Logan et al., 2000), *C. hominis* from humans (Lawson et al., 2001), *C. hyolei* was transferred to *C. coli* (Vandamme and On, 2001), *C. insulaenigrae* was isolated from seals and porpoise (Foster et al., 2004), *C. canadensis* from wild birds (Inglis et al., 2007), *C. avium* from poultry (Rossi et al., 2009), *C. cuniculorum* from rabbits (Zanoni et al., 2009), and *C. peloridis* was isolated from humans and mollusks (Debruyne et al., 2009).

Recently, between 2010 and 2015, six new species have been added to the genus *Campylobacter*; *B. ureolyticus* was finally transferred to *Campylobacter* as *C. ureolyticus* (Vandamme et al., 2010), *C. subantarcticus* was isolated from wild birds in the sub-Antarctic region (Debruyne et al., 2010b), *C. volucris* from black-headed gulls (Debruyne et al., 2010a), *C. troglodytes* was isolated from chimpanzees (Kaur et al., 2011), *C. corcagensis* from captive lion-tailed macaques (Koziel et al., 2014), and *C. iquaniorum* from reptiles (Gilbert et al., 2015).

3.3 CURRENT TAXONOMIC UPDATE ON SPECIES IN THE GENUS *CAMPYLOBACTER*

At present, the genus *Campylobacter* contains 27 species and 8 subspecies (Table 3.1). The current species and subspecies in the genus *Campylobacter* are highlighted below.

3.3.1 *CAMPYLOBACTER FETUS* SUBSP. *FETUS*, AND *CAMPYLOBACTER FETUS* SUBSP. *VENEREALIS*

C. fetus is reputed to be the first species to be isolated among the *Campylobacter* group (McFadyean and Stockman, 1913). *C. fetus* subsp. *fetus* is associated with abortion in sheep and cattle, and has also been isolated from a wide range of sources, such as chickens, reptiles, and humans. On the other hand, *C. fetus* subsp. *venerealis* is the causative agent of bovine genital campylobacteriosis that causes infertility, abortion, and embryonic death in cows. Differentiation between the two subspecies is based on a few phenotypic tests such as tolerance to 1% glycine, and pathogenic differences. Several molecular techniques have been used to discriminate between the two subspecies—for example, AFLP (Duim et al., 2001), RAPD-PCR (Tu et al., 2005), and PFGE (On and Harrington, 2001).

3.3.2 *CAMPYLOBACTER HYOINTESTINALIS*

This species consist of two subspecies: *C. hyointestinalis* subsp. *hyointestinalis*, and *C. hyointestinalis* subsp. *lawsonii*. *C. hyointestinalis* subsp. *hyointestinalis*, originally isolated from pigs, has been isolated from the intestine of hamsters, fecal samples from cattle, deer, and humans (Gebhart et al., 1985). It is associated with enteritis and diarrhea in these animals and humans, although the pathogenic role is not fully elucidated. *C. hyointestinalis* subsp. *lawsonii* occurs in the stomach of pigs. Its pathogenic role is not clear.

3.3.3 *CAMPYLOBACTER JEJUNI* AND *CAMPYLOBACTER COLI*

C. jejuni was originally isolated from the feces of cattle with diarrhea (Smith and Orcutt, 1927). *C. coli* were isolated from feces of pigs with diarrhea (Doyle, 1948).

Table 3.1 List of Species and Subspecies of *Campylobacter* and Their Common Hosts

Taxa	Host	References
C. fetus subsp. *fetus*	Cattle, sheep	Sebald and Veron (1963)
C. fetus subsp. *venerealis*	Cattle, sheep	Sebald and Veron (1963)
C. jejuni subsp. *jejuni*	Poultry, cattle, human	Jones et al. (1931)
C. jejuni subsp. *doylei*	Humans	Steele and Owen (1988)
C. coli	Pigs	Doyle (1948)
C. concisus	Humans	Tanner et al. (1981)
C. curvus	Humans	Tanner et al. (1984)
C. hyointestinalis subsp. *hyointestinalis*	Pigs, cattle, humans	Gebhart et al. (1985)
C. hyointestinalis subsp. *lawsonii*	Pigs	On et al. (1995)
C. lari subsp. *lari*	Seagulls, dogs, shellfish	Benjamin et al. (1983)
C. lari subsp. *concheus*	Humans, shellfish	Debruyne et al. (2009)
C. rectus	Humans	Vandamme et al. (1991)
C. upsaliensis	Cats, dogs, monkeys	Stanley et al. (1992)
C. helveticus	Cats, dogs	Stanley et al. (1992)
C. gracilis	Humans	Vandamme et al. (1995)
C. showae	Humans	Etoh et al. (1993, 1998)
C. sputorum	Cattle, pigs, humans	On et al. (1998)
C. lanienae	Pigs	Logan et al. (2000)
C. hominis	Humans	Lawson et al. (2001)
C. mucosalis	Pigs	Lawson et al. (2001)
C. insulaenigrae	Seals, porpoise	Foster et al. (2004)
C. canadensis	Wild birds	Inglis et al. (2007)
C. cuniculorum	Rabbits	Zanoni et al. (2009)
C. peloridis	Humans, mollusks	Debruyne et al. (2009)
C. avium	Poultry	Rossi et al. (2009)
C. ureolyticus	Humans	Vandamme et al. (2010)
C. volucris	Wild birds, humans	Debruyne et al. (2010a)
C. subantarcticus	Wild birds	Debruyne et al. (2010a)
C. troglodytis	Chimpanzees	Kaur et al. (2011)
C. corcagiensis	Captive lion-tailed macaques	Koziel et al. (2014)
C. iquaniorum	Lizards, chelonians	Gilbert et al. (2015)

C. jejuni consists of two subspecies; *C. jejuni* subsp. *jejuni* and *C. jejuni* subsp. *doylei*. *C. jejuni* subsp. *doylei* differs from *C. jejuni* subsp. *jejuni* by the absence of nitrate reduction, cephalothin susceptibility, and a weak catalase reaction. *C. jejuni* and *C. coli* are the most important human enteropathogens among the Campylobacters. *C. jejuni* and *C. coli* are closely related phenotypically. They differ only in the ability of *C. jejuni* to hydrolyze sodium hippurate, while *C. coli* are negative. However, hippurate-negative

C. jejuni have been reported. A wide range of genotypic methods has been used to differentiate *C. jejuni* from *C. coli* (Sails et al., 2001; Jensen et al., 2005).

3.3.4 *CAMPYLOBACTER LARI*

C. lari has been isolated from diverse sources, that is, intestinal contents of seagulls and other animals, river water, and shellfish. *C. lari* has also been isolated from cases of human diarrhea and bacteremia. In addition, it has been isolated from extraintestinal infections in immune-deficient patients. *C. lari* was originally referred to as nalidixic acid-resistant thermophilic *Campylobacter* (NARTC) group. They were differentiated from *C. jejuni* and *C. coli* mainly by their resistance to nalidixic acid and absence of indoxyl acetate hydrolysis. Endtz et al. (1997) and Duim et al. (2004) reported a striking heterogeneity among and within the different groups of *C. lari*, and identified four distinct subgroups. Later, nalidixic acid susceptible strains (NASC strains), urease producing strains (UPTC strains), and urease-producing nalidixic acid susceptible strains were identified as *C. lari* variants. Debruyne et al. (2009), after analyzing *C. lari*-like strains isolated from shellfish and humans using phenotypic, 16S rRNA, *hsp*60 gene sequence and DNA–DNA hybridization analysis, proposed a new subspecies for *C. lari* as *C. lari* subsp. *concheus* and *Campylobacter lari* subsp. *lari*.

3.3.5 *CAMPYLOBACTER SPUTORUM*

C. sputorum consists of three biovars (On et al., 1998). *C. sputorum* bv. *sputorum* produces neither catalase nor urease. They are found in the oral cavity, feces, and abscesses and other skin lesions of humans, and feces of sheep and pigs. *C. sputorum* bv. *fecalis* produces catalase but not urease, and has been isolated from the feces of sheep and cattle. *C. sputorum* bv. *paraureolyticus* produces urease but not catalase, and has been isolated from the feces of cattle and human diarrhea. The pathogenicity of the three strains is unknown.

3.3.6 *CAMPYLOBACTER MUCOSALIS*

It was first isolated in 1981 from proliferative enteritis in pigs, and named as a subspecies of *C. sputorum*, *C. sputorum* subsp. *mucosalis* (Lawson et al., 1981). DNA–DNA hybridization demonstrated that it was a different *C.* species (Roop et al., 1985), and was confirmed as such by Lawson et al. (2001). It has been isolated from the tongue of patients with oral malodor (Tyrrell et al., 2003). The clinical significance of this strain is not known.

3.3.7 *CAMPYLOBACTER CONCISUS, CAMPYLOBACTER SHOWAE, CAMPYLOBACTER CURVUS, CAMPYLOBACTER RECTUS, AND CAMPYLOBACTER GRACILIS*

They are all hydrogen-requiring *Campylobacter* species. *C. concisus* and *C. curvus* are isolated from the gingival crevices of humans with gingivitis and periodontitis

(Tanner et al., 1981, 1984). *C. concisus* has also been isolated from the stool and blood samples of children and adults. DNA–DNA hybridization, RAPD-PCR, PFGE, and whole protein electrophoresis have revealed that *C. concisus* are heterogenous (Engberg et al., 2005; Vandamme et al., 1989; Matsheka et al., 2002). *C. curvus* has been implicated in hepatic and bronchial abscesses. *C. rectus* has been isolated from gingival crevices of humans' appendicitis, patients with Barrett's esophagus, and extraoral abscesses. *C. gracilis* has been isolated from gingival crevices, soft tissue abscesses, pneumonia, and empyema in humans. *C. showae* has been isolated from human dental plaque, infected root canals, and periodontal lesions.

3.3.8 *CAMPYLOBACTER UPSALIENSIS* AND *CAMPYLOBACTER HELVETICUS*

C. upsaliensis, isolated in 1991 from fecal samples of healthy and diarrheic dogs (Sandstedt and Ursing, 1991), is catalase-negative or weakly positive, and has also been isolated from human in abortion cases, bacteremia, abscesses, and gastroenteritis. *C. helveticus* has been isolated from feline and canine fecal samples. At the present time, no strains have been isolated from humans, yet. *C. upsaliensis* is implicated as a human enteric pathogen because of its high isolation rate in diarrheic patients worldwide.

3.3.9 *CAMPYLOBACTER HOMINIS*

Lawson et al. (1998) described it as "Candidatus hominis" (candidate species) isolated from the gastrointestinal tract of humans. The species has now been successfully cultured and characterized (Lawson et al., 2001), and has been isolated from both healthy and diarrheic human fecal samples.

3.3.10 *CAMPYLOBACTER LANIENAE*

C. lanienae was first isolated from fecal samples of two abattoir workers (Logan et al., 2000) during a routine screening of healthy abattoir workers. It was later isolated from the fecal samples of healthy pigs, and bovine fecal samples. PCR assay, 16S rRNA gene sequence, and DNA homology confirmed it as a new species.

3.3.11 *CAMPYLOBACTER INSULAENIGRAE*

Phenotypic and genotypic analyses were performed on four *Campylobacter*-like organisms isolated from marine mammals (three common seals and a porpoise) in Northern Scotland. 16S rRNA gene sequencing and DNA–DNA hybridization studies confirmed the bacteria as belonging to distinct new *Campylobacter* species, *C. insulaenigrae* (Foster et al., 2004). *C. insulaenigrae* has also been isolated from northern elephant seals in California, USA. The pathogenic potential of *C. insulaenigrae* is unknown.

3.3.12 *CAMPYLOBACTER CANADENSIS*

Ten isolates of an unknown *Campylobacter* species isolated from cloacal swabs obtained from captured adult whooping cranes (*Grus americana*) in Canada were identified as *Campylobacter*, based on generic PCR and 23S rRNA gene sequence. The 10 isolates formed a robust clade that was distinct from known *Campylobacter* species based on 16S rRNA, *rpo*B, and cpn6O gene sequences. The novel *Campylobacter* species was named *C. canadensis* (Inglis et al., 2007). The pathogenic potential of *C. canadensis* is unknown.

3.3.13 *CAMPYLOBACTER CUNICULORUM*

Eight strains of an unknown *Campylobacter* species were isolated from the cecal contents of rabbits (*Oryctalagu cuniculus*). Genus-specific PCR initially identified them as *Campylobacter*, but none were identified using species-specific PCR for known thermophilic *Campylobacter* species. Phylogenetic analysis based on 16S rRNA gene, *rpo*B, and *gro*EL sequences revealed that all the strains formed a robust clade that was distinct from known *Campylobacter* species. It was closely related to *C. helveticus*. The novel species was named *C. cuniculorum* (Zanoni et al., 2009).

3.3.14 *CAMPYLOBACTER AVIUM*

Three strains of an unusual hippurate-positive *Campylobacter* species isolated from cecal contents of broiler chickens and a turkey analyzed by 16S rRNA, *rpo*B, and *gro*El gene sequences revealed that these strains formed a distinct clade from other *Campylobacter* species. Amplified fragment length polymorphism (AFLP) and whole cell protein electrophoresis confirmed them as a different *Campylobacter* species, named *C. avium* (Rossi et al., 2009).

3.3.15 *CAMPYLOBACTER PELORIDIS*

A polyphasic study was undertaken to clarify the taxonomic position of *C. lari*-like strains isolated from shellfish and humans. Fluorescent AFLP and whole cell protein electrophoresis analysis revealed existence of two clades different from *C. lari* and other *Campylobacter* species. Phenotypic analysis, 16S rRNA, *hsp*60 gene sequence analysis, and DNA–DNA hybridization studies further confirmed that 10 strains represented a novel *Campylobacter* species, named *C. peloridis* (Debruyne et al., 2009).

3.3.16 *CAMPYLOBACTER UREOLYTICUS*

The protein profiles, genomic AFLP patterns and 16S rRNA, and *cpn*60 gene sequences of a diverse collection of 26 *B. ureolyticus* were used to reassess the taxonomy of these bacterial species. Based on the results, *B. ureolyticus* was reclassified as *C. ureolyticus* (Vandamme et al., 2010). Strains have been isolated from superficial ulcers, soft tissue infections, nongonococcal, nonchlamydial urethritis, and periodontal disease.

3.3.17 *CAMPYLOBACTER VOLUCRIS*

Three *C. lari*-like strains were isolated during a study on the prevalence of *C. jejuni* in black-headed gulls (*Larus ridibundies*) in Sweden. Characterization of the isolates by AFLP and whole-cell protein SDS-PAGE revealed that they formed a distinct group in the genus *Campylobacter*, and were named *C. volucris* (Debruyne et al., 2010a).

3.3.18 *CAMPYLOBACTER SUBANTARCTICUS*

Six *C. lari*-like strains isolated from wild birds in the sub-Antarctic region were characterized by whole-cell protein electrophoresis and ALFP analysis that revealed they were distinct from *C. lari* and all other known *Campylobacter* species. 16S rRNA and DNA–DNA hybridization confirmed them as new species, named *C. subantarcticus* (Debruyne et al., 2010b). Their pathogenicity is unknown.

3.3.19 *CAMPYLOBACTER TROGLODYTES*

Nineteen *Campylobacter* strains were isolated from the feces of human-habituated chimpanzees (*Pan troglodytes schweinfurthii*) in Tanzania. The isolates were virtually identical (with a single nucleotide base difference), using 16S rRNA gene sequencing, and were most closely related to *C. helveticus* and *C. upsaliensis*. Whole-cell protein electrophoresis, AFLP analysis, 16S rRNA, and *hsp*60 sequence analysis and DNA–DNA hybridization studies confirmed that the species represented a distinct genomic species, and was named *C. troglodytes* (Kaur et al., 2011). Further studies are required to determine the pathogenicity of the species, in chimpanzees and humans.

3.3.20 *CAMPYLOBACTER CORCAGIENSIS*

Recently, an investigation of the prevalence of *C. ureolyticus* in a variety of animals, including the feces of captive lion-tailed macaques (*Macaca silenus*) identified them as *C. ureolyticus*. Three strains obtained from captive lion-tailed macaques, based on colony morphology, formed a distinct clade within the genus *Campylobacter*, based on their 16S rRNA and *hsp*60 sequences, and MALDI-TOF profiles. The unique species was further supported by phenotypic characteristics of the isolates. The new species was named *C. corcagiensis* (Koziel et al., 2014). The pathogenicity of the species has not been established yet.

3.3.21 *CAMPYLOBACTER IQUANIORUM*

In a recent study to isolate members of the class Epsilon proteobacteria from reptiles, five *Campylobacter* strains were isolated from reptiles (lizards and chelonians). Initial AFLP, PCR, and 16S rRNA sequence analysis showed that these strains were most closely related to *C. fetus* and *C. hyointestinalis*. The strains were characterized by 16S rRNA and *atp*A gene sequence analysis, MALDI-TOF mass spectrometry, and conventional phenotypic testing. Whole genome sequences were determined for two strains, and the average nucleotide and amino acid identities were determined for the two strains. The strains formed a robust phylogenetic clade distinct from all known members of the genus *Campylobacter*, and were named *C. iquaniorom* (Gilbert et al., 2015).

3.4 PHYSIOLOGICAL CHARACTERISTICS OF *CAMPYLOBACTER*

The genus *Campylobacter* are generally Gram-negative, nonsaccharoylytic bacteria with microaerobic growth requirements, and a low G + C content. Energy is obtained from amino acids or tricarboxylic acid cycle intermediates. They do not ferment or oxidize carbohydrates. The cells are usually curved, S-shaped, or spiral rods that are 0.2–0.8 μm-wide by 0.5–5 μm-long. Cells in old cultures, or exposed to oxygen, form spherical or coccoid bodies, and cells of most species are motile.

The physiological characteristics that serve to differentiate the species within the genus *Campylobacter* are discussed for some species, and the characteristics for all the species are presented in Table 3.2.

3.4.1 *CAMPYLOBACTER FETUS* SUBSP. *FETUS*, AND *CAMPYLOBACTER FETUS* SUBSP. *VENEREALIS*

Strains grow optimally at 37°C, can grow at 25°C, but not at 42°C, under microaerobic conditions. Strains are oxidase- and catalase-positive, and do not hydrolyze indoxyl acetate or hippurate. They are sensitive to cephalothin and resistant to nalidixic acid. *C. fetus* subsp. *fetus* grow in medium containing 1% glycine, while *C. fetus* subsp. *venerealis* do not. Tolerance to 0.1% potassium permanganate, 0.0055 basic fuchsin, and 64 mg/L cefoperazone may discriminate between the two subspecies (On and Harrington, 2001).

3.4.2 *CAMPYLOBACTER CORCAGIENSIS*

C. corcagiensis are oxidase- and urease-positive. They are alkaline phosphatase-positive, some strains reduce nitrates, and some hydrolyze indoxyl acetate (Koziel et al., 2014). *C. corcagiensis* do not hydrolyze hippurate and gelatin, however all strains produce H_2S on TSI and SIM media. Strains grow on blood agar at 25, 37, and 42°C, under anaerobic conditions, but not at 15°C. Strains also grow on blood agar supplemented with nalidixic acid, 1% glycine, 20% NaCl, and 1% bile. Two types of colonies are observed: small pinpoint colonies (1 mm diameter), and flat spreading colonies (1–3 mm diameter). They do not grow on media containing 2, 3, 5-triphenyltetrazolium chloride and cefoperazone. Strains do not grow on MacConkey agar, and growth on unsupplemented nutrient agar is weak.

Table 3.2 Physiological Characteristics of *Campylobacter* Species

Characteristic	1	2	3	4	5	6	7	8	9	10
α-Hemolysis	2	+	NA	NA	NA	V	(2)	(2)	(2)	2
Oxidase	+	+	+	+	+	+	+	V	+	+
Catalase	V	+	+	+	+	(2)	+	2	2	+
Hippurate hydrolysis	2	−	−	−	(−)	2	2	2	(2)	2
Urease production	V	−	NA	+	+	+	2	2	2	2
Nitrate reduction	V	+	+	(+)	(−)	+	+	(2)	+	+
Alkaline phosphatase	2	−	NA	+	+	2	2	V	V	2
Production on TSI agar of:										
$\quad$ H_2S	V	−	NA	+	NA	2	V	2	(2)	2
$\quad$ Acid and H_2S	(2)	NA	NA	NA	NA	2	2	2	2	2
Indoxyl acetate hydrolysis	2	+	−	V	(−)	2	+	2	V	2
Growth at/in/on:										
$\quad$ 25°C (microaerobic)	2	−	−	+	−	2	2	2	2	+
$\quad$ 30°C (microaerobic)	2	NA	NA	+	NA	+	+	(+)	+	+
$\quad$ 37°C (microaerobic)	+	+	NA	+	+	2	+	+	V	+
$\quad$ 42°C (microaerobic)	+	(+)	+	+	+	V	+	(+)	V	(+)
$\quad$ 37°C (anaerobic)	+	−	NA	−	NA	+	2	+	+	(2)
$\quad$ Ambient O_2	2	−	NA	NA	NA	2	2	2	2	2
$\quad$ Nutrient agar	2	+	−	w	NA	+	+	(2)	+	+
$\quad$ CCDA	+	(+)	NA	NA	NA	V	+	(2)	(+)	+
$\quad$ MacConkey agar	+	−	w	−	NA	(2)	V	2	(+)	(+)
$\quad$ 1% glycine	V	−	−	+	+	+	+	(2)	+	+
$\quad$ 3.5% NaCl	2	−	−	+	+	+	2	2	2	2
Resist to nalidixic acid	V	V	+	+	NA	2	2	(+)	+	+
Resist to cephalothin	2	(+)	+	NA	+	2	+	2	2	2
Requirement for H_2	2	−	NA	NA	NA	NA	2	+	+	2
Nitrite reduction	+	NA	NA	NA	NA	NA	2	2	NA	2
Flagella	+	NA	+	+	+	NA	+	+	+	+

Characteristic	11	12	13	14	15	16	17	18	19	20
α-Hemolysis	V	2	+	2	V	V	NA	+	+	2
Oxidase	+	2	+	+	+	+	+	+	+	+
Catalase	(+)	V	2	2	+	+	+	V	+	+
Hippurate hydrolysis	2	2	2	2	2	2	2	+	+	2
Urease production	2	2	2	2	2	2	2	2	2	2
Nitrate reduction	+	(+)	+	V	+	+	+	2	+	+
Alkaline phosphatase	2	2	2	2	2	(2)	NA	2	2	+
Production on TSI agar of:										
$\quad$ H2S	2	2	2	2	⊦	+	2	?	2	2
$\quad$ Acid and H_2S	2	2	2	NA	2	2	NA	2	2	V
Indoxyl acetate hydrolysis	2	V	+	2	2	2	2	+	+	2

(Continued)

Table 3.2 Physiological Characteristics of *Campylobacter* Species (*cont.*)

Characteristic	11	12	13	14	15	16	17	18	19	20
Growth at/in/on:										
25°C (microaerobic)	+	2	2	2	(2)	2	2	2	2	2
30°C (microaerobic)	+	V	V	2	+	+	NA	+	+	NA
37°C (microaerobic)	+	2	+	2	+	+	NA	+	+	+
42°C (microaerobic)	2	V	+	NA	+	+	2	2	+	+
37°C (anaerobic)	V	+	2	+	2	+	2	2	2	+
Ambient O_2	2	2	2	2	2	2	2	2	2	2
Nutrient agar	+	+	(+)	NA	+	+	NA	+	+	NA
CCDA	+	V	+	NA	+	+	NA	+	+	NA
MacConkey agar	V	(+)	2	2	V	V	NA	2	2	+
1% glycine	2	+	V	+	+	V	+	(2)	+	2
3.5% NaCl	2	2	2	+	2	2	2	2	2	2
Resist to nalidixic acid	V	V	2	V	+	+	+	2	2	+
Resist to cephalothin	2	2	2	2	(2)	2	+	2	+	+
Requirement for H_2	2	NA	2	NA	V	V	NA	2	2	2
Nitrite reduction'	NA	NA	2	NA	2	2	2	NA	2	V
Flagella	+	2	+	2	+	+	NA	+	+	+

Characteristic	21	22	23	24	25	26	27	28	29
α-Hemolysis	V	2	+	+	+	+	NA	–	NA
Oxidase	+	+	+	V	+	+	+	+	+
Catalase	+	2	(2)	+	V	2	+	v	+
Hippurate hydrolysis	2	2	2	2	2	2	–	+	–
Urease production	NA	V	2	2	2^a	2	2	NA	–
Nitrate reduction	+	2	+	+	+	+	NA	+	NA
Alkaline phosphatase	2	(+)	2	2	2	2	NA	–	NA
Production on TSI agar of:									
H_2S	2	+	2	V	+	2	NA	–	–
Acid and H_2S	2	2	2	2	2	2	NA	NA	
Indoxyl acetate hydrolysis	2	2	+	V	2	+	NA	+	NA
Growth at/in/on:									
25°C (microaerobic)	2	2	2	2	2	2	NA	–	–
30°C (microaerobic)	+	+	V	+	(+)	+	–	–	–
37°C (microaerobic)	+	+	2	V	+	+	+	+	NA
42°C (microaerobic)	+	+	(2)	V	+	(+)	+	+	+
37°C (anaerobic)	–	2	+	+	+	+	2	NA	–
Ambient O_2	2	2	2	2	2	2	NA	NA	NA
Nutrient agar	+	+	(2)	V	+	+	+	–	–
CCDA	+	+	2	+	(+)	+	+	–	NA
MacConkey agar	2	(+)	2	+	V	2	NA	–	(–)
1% glycine	+	V	+	V	+	+	+	–	(+)
3.5% NaCl	2	2	2	2	V	2	+	–	

Table 3.2 Physiological Characteristics of *Campylobacter* Species (*cont.*)

Characteristic	21	22	23	24	25	26	27	28	29
Resist to nalidixic acid	V	(+)	(+)	2	(+)	2	(+)	–	+
Resist to cephalothin	+	2	2	2	2	(2)	(–)	+	–
Requirement for H_2	2	+	+	+	+	2	NA	V	–
Nitrite reduction	2	NA	NA	NA	2	2	NA	NA	NA
Flagella	+	+	+	+	+	+	NA	NA	NA

The references for this table are included in the respective paragraphs of the main text
Taxa: 1, C. canadensis; 2, C. cuniculorum; 3, C. volucris; C. corcagiensis; 5, C. troglodytes; 6,
C. ureolyticus; 7, C. coli; 8, C. concisus; 9, C. curvus; 10, C. fetus subsp. fetus; 11, C. fetus
subsp. venerealis; 12, C. gracilis; 13, C. helveticus; 14, C. hominis; 15, C. hyointestinalis subsp.
hyointestinalis; 16, C. hyointestinalis subsp. lawsonii; 17, C. insulaenigrae; 18, C. jejuni subsp. doylei;
19, C.jejuni subsp. jejuni; 20, C. lanienae; 21, C. lari; 22, C. mucosalis; 23, C. rectus; 24, C. showae;
25, C. sputorum; 26, C. upsaliensis; 27, C. peloridis; 28, C avium; 29, C. subantarcticus. (+), 90–
100% of strains positive; (+), 75–89% of strains positive; V, 26–74% of strains positive; (2), 11–25% of
strains positive; (–), 0–10% of strains positive; (–), negative; w, weak; NA, No data available
[a]Strains of C. sputorum bv. paraureolyticus are urease-positive; other strains are urease-negative
(On et al., 1998)

3.4.3 *CAMPYLOBACTER AVIUM*

Strains grow at 37 and 42°C under microaerobic conditions, but not at 25°C, or under anaerobic or aerobic conditions. Colonies appear non α-hemolytic, flat, grayish, and finely granular, with an irregular edge, and show a tendency to spread along the line of streak, and to swarm and coalesce (Rossi et al., 2009). Most strains do not require hydrogen to grow. Strains are oxidase-positive and weak catalase, urease, and alkaline phosphatase-negative, and hydrolyze hippurate and indoxyl acetate. They reduce nitrate but not selenite or triphenyl tetrazolium chloride. Strains do not produce H_2S on TSI agar. Most strains grow in the presence of 1% bile, and do not grow on nutrient agar without blood, MacConkey agar, or in the presence of 1% (w/v) glycine or 2% (w/v) NaCl. They grow on CCDA (after 4–5 days of incubation). Strains are susceptible to nalidixic acid (30 mg) and resistant to cephalothin (30 mg).

3.4.4 *CAMPYLOBACTER TROGLODYTES*

Strains grow at 37 and 42°C under microaerobic conditions. Strains are oxidase, catalase, and alkaline phosphatase-positive. They grow on 1% glycine and are sensitive to nalidixic acid. Most of the strains are positive for growth on triphenyl tetrazolium chloride, and negative for selenite reduction. They do not grow on 2% NaCl and 2 % bile. Most of the isolates are negative to indoxyl acetate and nitrate reduction. They are resistant to caphalothin, and most of the isolates do not hydrolyze hippurate (Kaur et al., 2011).

3.4.5 *CAMPYLOBACTER INSULAENIGRAE*

Strains grow at 37°C under microaerobic conditions. Growth does not occur at 25 or 42°C, or in aerobic or anaerobic conditions. Growth also occurs in the presence of 1% glycine, but not in the presence of 3.5% NaCl. They are oxidase- and catalase-positive. They are urease-negative, reduced nitrate, but nitrites are not reduced. Strains produce H_2S on TSI agar. They do not hydrolyze hippurate and indoxyl acetate (Foster et al., 2004). They are resistant to both naliudixic acid (30 mg) and cephalothin (30 mg).

3.4.6 *CAMPYLOBACTER LANIENAE*

They grow at 37 and 42°C, but not at 25°C, under microaerobic conditions for 2–3 days. Colonies are smooth, entire, translucent, and cause some greening of blood agar (Logan et al., 2000), and grow weakly under anaerobic condition. Urease, DNase, arylsulfatase, and pyrazinamidase are not produced. Nitrate and nitrite are reduced. H_2S is not produced on TSI, indoxyl acetate and hippurate are not hydrolyzed. They do not grow in 1% glycine or 2% NaCl, and are resistant to nalidixic acid and cephalothin.

3.4.7 *CAMPYLOBACTER UREOLYTICUS*

The physiological description of *C. ureolyticus* is mainly based on the data for *B. ureolyticus* (On and Holmes 1995; On et al., 1996; Vandamme et al., 2005). Strains grow on blood agar. Optimal growth occurs in hydrogen enriched in aerobic conditions. They do not grow microaerobically on common agar bases in an atmosphere without hydrogen. They will not grow in air, in a CO_2 enriched atmosphere, or in an atmosphere containing 5% O_2, 10% CO_2, and 85% N_2 on common agar bases. Anaerobic growth occurs with formate and fumarate in the medium. They grow at 30 and 37°C on media containing 0.1% trimethylamine-*N*-oxide, 1% glycine, 0.05% sodium fluoride, 0.032% methyl orange, or 2–4% NaCl. No growth occurs in the presence of 1–2% bile, 0.04%, 2,3,5-triphenyl tetrazolium chloride, or 0.05% basic fuchsin. Most strains grow on nutrient agar and buffered charcoal yeast medium. Growth on CCDA and MacConkey agar is strain-dependent. Oxidase and urease activity is present, no hydrolysis of hippurate, DNase activity, alkaline phosphatase, reduction of triphenyl-tetrazolium chloride, or production of H_2S on TSI. Gelatinase activity is present (Jackson and Goodman, 1978). Nitrate is reduced, but not selenite. α-Hemolysis, catalase, and indoxyl acetate hydrolysis are strain-dependent. No hydrolysis of casein or growth on casein medium occurs. They are susceptible to nalidixic acid and cephalothin.

3.4.8 *CAMPYLOBACTER PELORIDIS*

Strains grow on 5% blood agar under microaerobic conditions. They are oxidase- and catalase-positive. Growth occurs on media containing 1% glycine and 4 mg/L metronidazole. Most strains (80%) grow on media containing 2% NaCl or 32 mg nalidixic acid. Most strains (90%) do not grow on media containing 32 mg cephalothin. Strains do not grow on media containing 0.05% safranin or 32 mg carbenicilin (Debruyne et al., 2009).

3.4.9 *CAMPYLOBACTER CUNICULORUM*

Strains grow on nutrient agar and blood agar under microaerobic conditions, colonies are α-hemolytic on blood agar. Colony appearance on MCCDA and cefoperazone, amphotericin B, teicoplanin, is similar to that on NA. They are strictly microaerobic, grow at 37 and 42°C, but not at 25°C or under anaerobic or aerobic conditions. H_2 is not required for growth (Zanoni et al., 2009). They are oxidase- and catalase-positive, urease-negative, γ-glutamyltranspeptidase- and alkaline phosphatase-negative. They hydrolyze indoxyl acetate but not hippurate, and reduce nitrate but not selenite. Some strains reduce triphenyltetrazolium chloride. They do not grow on MacConkey agar or in the presence of 1% glycine and 2% NaCl, and only few strains grow in the presence of 1% bile. Most strains are resistant to nalidixic acid (30 mg) and cephalothin (30 mg).

3.4.10 *CAMPYLOBACTER VOLUCRIS*

Strains produce oxidase and catalase, and reduce nitrate and selenite. They do not hydrolyze indoxyl acetate and hippurate, H_2 is not produced in TSI agar, and are alkaline phosphatase-negative. They grow on blood agar at 37 and 42°C, but not at 25°C or room temperature (22°C), under microaerobic and anaerobic (scanty growth) conditions, and also not under aerobic conditions. Strains also grow on blood agar medium containing 32 mg nalidixic acid, 32 mg cephalothin, 4 mg metronidazole, and 32 mg carbenicilin, and grow weakly on MacConkey and starch agar media as well as on media containing 0.1% ox bile and 0.032% methyl orange. Strains do not grow on 0.04% triphenyl tetrazolium chloride (TTC) medium and on unsupplemented nutrient agar, on a minimum media on caisein, lecithin, or tryrosine media. No growth on media containing 2% NaCl, 1% glycine, 0.02% safranin, 0.001% sodium arsenite, 0.1% potassium permanganate, 0.005% basic fuchsin, 0.0005% crystal violet, 0.1% janus green, 0.02% pyronin, or 64 mg cefoperazone (Debruyne et al., 2010a).

3.4.11 *CAMPYLOBACTER SUBANTARCTICUS*

Strains are oxidase- and catalase-positive, do not hydrolyze indoxyl acetate or hippurate and do not produce H_2S on TSI agar. Strains do not produce selenite, and grow at 37 and 42°C, but not at room temperature (22°C) or 25°C under microaerobic conditions. Strains do not grow under aerobic conditions at 37 or 25°C. They grow at 37°C under anaerobic conditions on unsupplemented blood agar, or on blood agar supplemented with 0.1% trimethylamine-*N*-oxide; strains also grow on media containing 1–2% ox bile, 2% NaCl, 32 mg nalidixic acid, 100 U, 5-fluorouracil, and 0.05% sodium fluoride. Most strains grow on media containing 1% glycine. Most strains do not grow on media containing 4 mg metronidazole or on MacConkey agar. Strains do not grow on unsupplemented nutrient agar, casein agar, lecithin agar, tyrosine agar, or on media containing 0.02 0.05% safranin, 0.1% sodium deoxycholate, 0.1% potassium permanganate, 0.005% crystal violet, 3.5% NaCl, 32 mg cephalothin, or 32 mg carbenicillin, α-hemolysis on 5% blood agar (Debruyne et al., 2010b).

3.4.12 *CAMPYLOBACTER CANADENSIS*

Strains grow on CCDA, Columbia agar, and trypticase soy agar similar to the growth on Karmali agar. Strains are not α-hemolytic, grow at 35, 37, 40, and 42°C, but not at 25°C under microaerobic conditions. All strains also grow under anaerobic conditions; all strains do not hydrolyze indoxyl acetate or hippurate, produce alkaline phosphatase, or require H_2 to grow. All strains produce oxidase, grow on MacConkey agar, and reduce nitrite. No strains grow on nutrient agar, some strains reduce nitrate (50%), produce catalase (40%), urease (50%) and L-glutamyl transpeptidase (80%), grow on 1% glycine (30%) and 3.5% NaCl (10%). Some strains are resistant to nalidixic acid (40%) and produce acid (20%) or H_2S (40%) on TSI agar. All strains are susceptible to cephalothin. No growth occurs in the presence of 2 or 3.5% NaCl at 40°C (Inglis et al., 2007).

REFERENCES

Benjamin, J., Leaper, S., Owen, R.J., Skirrow, M.B., 1983. Description of *Campylobacter laridis*, a new species comprising the nalidixic acid resistant thermophilic *Campylobacter* (NARTC) Group. Curr. Microbiol. 8, 231–238.

Bryner, J.H., Littleton, J., Gates, C., Kirkbride, C. A., Ritchie, A. E., 1986. *Flexispira rappini* gen. nov., sp. nov., a Gram negative rod from mammalian fetus and feces. Microbe, 86, Abstract XIV International Congress of Microbiology, Manchester, England, pp. 11–18.

Butzler, J.P., Dekeyser, P., Detrain, M., Dehaen, F., 1973. Related vibrio in stools. J. Pediatr. 82, 493–495.

Debruyne, L., Gevers, D., Vandamme, P., 2008. Taxonomy of the family Campylobacteraceae. In: Nachamkin, I., Szymanski, C.M., Blaser, M.J. (Eds.), Campylobacter. American Society for Microbiology, Washington, DC, pp. 3–26.

Debruyne, L., On, S.L., De Brandt, E., Vandamme, P., 2009. Novel *Campylobacter lari*-like bacteria from humans and molluscs: description of *Campylobacter peloridis* sp. nov., *Campylobacter lari* subsp. *concheus* subsp. nov. and *Campylobacter lari* subsp. *lari* subsp. nov. Int. J. Syst. Evol. Microbiol. 59, 1126–1132.

Debruyne, L., Broman, T., Bergstrom, S., Olsen, B., On, S.L., Vandamme, P., 2010a. *Campylobacter volucris* sp. nov., isolated from black-headed gulls (*Larus ridibundus*). Int. J. Syst. Evol. Microbiol. 60, 1870–1875.

Debruyne, L., Broman, T., Bergstrom, S., Olsen, B., On, S.L.W., Vandamme, P., 2010b. *Campylobacter subantarcticus* sp. nov., isolated from birds in the sub-Antarctic region. Int. J. Syst. Evol. Microbiol. 60, 815–819.

Dekeyser, P., Gossuin-Detrain, M., Butzler, J.P., Sternon, J., 1972. Acute enteritis due to related vibrios: first positive stool cultures. J. Infect. Dis. 125, 390–392.

Doyle, L.P., 1948. The etiology of swine dysentery. Am. J. Vet. Res. 9, 50–51.

Duim, B., Vandamme, P.A., Rigter, A., Laevens, S., Dijkstra, J.R., Wagenaar, J.A., 2001. Distribution of *Campylobacter* species by AFLP fingerprinting. Microbiology 147, 2729–2737.

Duim, B., Wagenaar, J.A., Dijkstra, J.R., Goris, J., Endtz, H.P., Vandamme, P.A., 2004. Identification of distinct *Campylobacter lari* genotypes by Amplified Fragment Length Polymorphism and protein electrophoresis profiles. Appl. Environ. Microbiol. 70, 18–24.

Endtz, H.P., Vliegenthart, J.S., Vandamme, P., Weverink, H.W., van den Braak, N.P., Verbrugh, H.A., van Belkum, A., 1997. Genotypic diversity of *Campylobacter lari* isolated from mussels and oysters in The Netherlands. Int. J. Food Microbiol. 34, 79–88.

Engberg, J., Bang, D.D., Aabenhus, R., Aarestrup, F.M., Fussing, V., Gerner–Smidt, P., 2005. Campylobacter concisus: an evaluation of certain phenotypic and genotypic characteristics. Clin. Microbiol. Infect. 11, 288–295.

Etoh, Y., Dewhirst, F.E., Paster, B.J., Yamamoto, A., Goto, N., 1993. *Campylobacter showae* sp. nov., isolated from the human oral cavity. Int. J. Syst. Bacteriol. 43, 631–639.

Etoh, Y., Yamamoto, A., Goto, N., 1998. Intervening sequences in 16S rRNA genes of *Campylobacter* sp.: diversity of nucleotide sequences and uniformity of location. Microbiol. Immun. 42, 241–243.

Florent, A., 1953. Isolement d'un vibrion saprophyte du sperme du taureau et du vagin de la vache (*Vibrio bubulus*) C. R. Soc. Biol. 147, 2066–2069.

Foster, G., Holmes, B., Steigerwalt, A.G., Lawson, P.A., Thorne, P., Byrer, D.E., Ross, H.M., Xerry, J., Thompson, P.M., Collins, M.D., 2004. *Campylobacter insulaenigrae* sp. nov., isolated from marine mammals. Int. J. Syst. Evol. Microbiol. 54, 2369–2373.

Fox, J.G., Chilvers, T., Goodwin, C.S., Taylor, N.S., Edmonds, P., Sly, L.I., Brenner, D.J., 1989. *Helicobacter mustelae* a new species resulting from the elevation of *Campylobacter pylori* subsp. *mustelae* to species status. Int. J. Syst. Bacteriol. 39, 301–303.

Gebhart, C.J., Edmonds, P., Ward, G.E., Kurtz, H.J., Brenner, D.J., 1985. *Campylobacter hyointestinalis* sp. nov.: a new species of *Campylobacter* found in the intestines of pigs and other animals. J. Clin. Microbiol. 21, 715–720.

Gilbert, M.J., Kik, M., Miller, W.G., Duim, B., Wagenaar, J.A., 2015. Campylobacter iguaniorum sp. nov., isolated from reptiles. Int. J. Syst. Evol. Microbiol. 65, 975–982.

Goodwin, C.S., Armstrong, J.A., Chilvers, T., Peters, M., Collins, M.D., Sly, L., McConnell, W., Harper, W.E.S., 1989. Transfer of *Campylobacter pylori* and *Campylobacter mustelae* to *Helicobacter* gen. nov. as *Helicobacter pylori* comb. nov. and *Helicobacter mustelae* comb. nov., respectively. Int. J. Syst. Bacteriol. 39, 397–405.

Inglis, G.D., Hoar, B.M., Whiteside, D.P., Morck, D.W., 2007. *Campylobacter canadensis* sp. nov., from captive whooping cranes in Canada. Int. J. Syst. Evol. Microbiol. 57, 2636–2644.

Jackson, F.L., Goodman, Y.E., 1978. *Bacteriodes ureolyticus*, a new species to accommodate strains previously identified as *Bacteriodes corrodens*, anaerobic. Int. J. Syst. Bacteriol. 28, 197–200.

Jensen, A.N., Andersen, M.T., Dalsgaard, A., Baggesen, D.L., Nielsen, E.M., 2005. Development of real-time PCR and hybridization methods for detection and identification of thermophilic *Campylobacter* species in pig fecal samples. J. Appl. Microbiol. 99, 292–300.

Jones, F.S., Orcutt, M., Little, R.B., 1931. Vibrios (*V. jejuni* n. sp.) associated with intestinal disorders in cows and calves. J. Exp. Microbiol. 81, 163–168.

Kaur, T., Singh, J., Huffman, M.A., Petrzelkova, K.J., Taylor, N.S., Xu, S., Dewhirst, F.E., Paster, B.J., Debruyne, L., Vandamme, P., Fox, J.G., 2011. *Campylobacter troglodytis* sp. nov., Isolated from Feces of Human-Habituated Wild Chimpanzees (*Pan troglodytes schweinfurthii*) in Tanzania. Appl. Environ. Microbiol. 77, 2366–2373.

King, E.O., 1957. Human infections with *Vibrio fetus* and a closely related vibrio. J. Infect. Dis. 101, 119–128.

Koziel, M., O'Doherty, P., Vandamme, P., Corcoran, G.D., Sleator, R.D., Lucey, B., 2014. *Campylobacter corcagiensis* sp. nov., isolated from feces of captive lion-tailed macaques (*Macaca silenus*). Int. J. Syst. Evol. Microbiol. 64, 2878–2883.

Lawson, G.H.K., Leaver, J.L., Pettigrew, G.W., Rowland, A.C., 1981. Some features of *Campylobacter sputorum* subspecies *mucosalis* subsp. nov., nom. rev. and their taxonomic significance. Int. J. Syst. Bacteriol. 31, 385–391.

Lawson, A.J., Linton, D., Stanley, J., 1998. 16S rRNA gene sequences of 'Candidatus *Campylobacter* hominis', a novel uncultivated species, are found in the gastrointestinal tract of healthy humans. Microbiology 144, 2063–2071.

Lawson, A.J., On, S.L.W., Logan, J.M.J., Stanley, J., 2001. *Campylobacter hominis* sp. nov. from the human gastrointestinal tract. Int. J. Syst. Evol. Microbiol. 51, 651–660.

Levi, A.J., 1946. A gastroenteritis outbreak probably due to a bovine strain of *Vibrio*. Yale J. Biol. Med. 18, 243.

Logan, J.M.J., Burnens, A.P., Linton, D., Lawson, A.J., Stanley, J., 2000. *Campylobacter lanienae* sp. nov., a new species isolated from workers in an abattoir. Int. J. Syst. Evol. Microbiol. 50, 865–872.

Marshall, B.J., Royce, H., Annear, D.I., Goodwin, C.S., Pearman, J.W., Warren, J.R., Armstrong, J.A., 1984. Original isolation of *Campylobacter pyloridis* from human gastric mucosa. Microbios Lett. 25, 83–88.

Matsheka, M.I., Elisha, B.G., Lastovica, A.L., On, S.L., 2002. Genetic heterogeneity of *Campylobacter concisus* determined by PFGE-based macrorestriction profiling. FEMS Microbiol. Lett. 211, 17–22.

McFadyean, J., Stockman, S., 1913. Abortion in sheep. Report of the Departmental Committee appointed to enquire into epizootic abortion. Part 3. Great Britain Board of Agriculture and Fisheries, London, pp. 12–23.

McLung, C.R., Patriquin, D.G., Davis, R.E., 1983. *Campylobacter nitrofigilis* sp. nov., a nitrogen-fixing bacterium associated with roots of *Spartina alterniflora* Loisel. Int. J. Syst. Bacteriol. 33, 605–612.

Neill, S.D., Campbell, J.N., O'Brien, J.J., Weatherup, S.T., Ellis, W.A., 1985. Taxonomic position of *Campylobacter cryaerophilia* sp. nov. Int. J. Syst. Bacteriol. 35, 342–356.

On, S.L., 2013. Isolation, identification and subtyping of *Campylobacter*: where to from here? J. Microbiol. Methods 95, 3–7.

On, S.L.W., Harrington, C.S., 2001. Evaluation of numerical analysis of PFGE-DNA profiles for differentiating *Campylobacter fetus* subspecies by comparison of phenotypic, PCR, PFGE and 16S rDNA sequencing methods. J. Appl. Microbiol. 90, 28–293.

On, S.L., Holmes, B., 1995. Classification and identification of campylobacters, helicobacters and allied taxa by numerical analysis of phenotypic characters. Syst. Appl. Microbiol. 18, 374–390.

On, S.L.W., Holmes, B., Sackin, M.J., 1996. A probability matrix for the identification of campylobacters, helicobacters and allied taxa. J. Appl. Bacteriol. 81, 425–432.

On, S.L.W., Atabay, H.I., Corry, J.E.L., Harrington, C.S., Vandamme, P., 1998. Emended description of *Campylobacter sputorum* and revision of its infrasubspecific (biovar) divisions, including *C. sputorum* bv. *paraureolyticus,* a urease-producing variant from cattle and humans. Int. J. Syst. Bacteriol. 48, 195–206.

On, S.L., Bloch, B., Holmes, B., Hoste, B., Vandamme, P., 1995. Campylobacter hyointestinalis subsp. lawsonii subsp. nov., isolated from the porcine stomach, and an emended description of Campylobacter hyointestinalis. Int. J. Syst. Bact. 45, 767–774.

Prèvot, A.R., 1940. Etudes de systèmatique bacterienne. V. Essai de classification des vibrions anaèrobies. Ann. Inst. Pasteur. 64, 117–125, (in French).

Roop, R.M., Smibert, H.R.M., Johnson, J.L., Krieg, N.R., 1985. *Campylobacter mucosalis* (Lawson, Learer, Pettigrew, and Rowland 1981) comb. nov.: emended description. Int. J. Syst. Bacteriol. 35, 189–192.

Rossi, M., Debruyne, L., Zanoni, R.G., Manfreda, G., Revez, J., Vandamme, P., 2009. *Campylobacter avium* sp. nov., a hippurate positive species isolated from poultry. Int. J. Syst. Evol. Microbiol. 59, 2364–2369.

Sails, A.D., Fox, A.J., Bolton, F.J., Wareing, D.R., Greenway, D.L., Borrow, R., 2001. Development of a PCR ELISA assay for the identification of *Campylobacter jejuni* and *Campylobacter coli*. Mol. Cell. Probes 15, 291–300.

Sandstedt, K., Ursing, J., 1991. Description of *Campylobacter upsaliensis* sp. nov, previously known as the CNW group. Syst. Appl. Microbiol. 14, 39–45.

Sebald, M., Veron, M., 1963. Teneur en bases da l'AND et classification des vibrions. Annales de L'institut Pasteur (Paris). 105, 897–910, (in French).

Skirrow, M.B., 1977. *Campylobacter* enteritis: a "new" disease. Br. Med. J. 2, 9–11.

Smith, T., Orcutt, M., 1927. Vibrios from calves and their serological relatives to *Vibrio fetus*. J Exp. Med. 45, 391–397.

Smith, T., Taylor, M.S., 1919. Some morphological and biological characters of the *Spirilla* (*Vibrio fetus*, n. sp.) associated with disease of the fetal membranes in cattle. J. Exp. Med. 30, 299–311.

Stanley, J., Burnens, A.P., Linton, D., On, S.L., Costas, M., Owen, R.J., 1992. *Campylobacter helveticus* sp. nov., a new thermophilic species from domestic animals: characterization, and cloning of a species-specific DNA probe. J. Gen. Microbiol. 138, 2293–2303.

Steele, T.W., Owen, R.J., 1988. *Campylobacter jejuni* subsp. *doylei* subsp. nov., a subspecies of nitrate-negative campylobacters isolated from human clinical specimens. J. Syst. Evol. Microbiol. 38, 316–318.

Stolz, J.F., Ellis, D.J., Blum, J.S., Ahmann, D., Lovley, D.R., Oremland, R.S., 1999. *Sulfurospirillum barnesii* sp. nov. & *Sulfurospirillum arsenophilum* sp. nov., new members of the *Sulfurospirillum* clade of the e Proteobacteria. Int. J. Syst. Bacteriol. 49, 1177–1180.

Tanner, A.C.R., Badger, S., Lai, C.H., Listgarten, M.A., Visconti, R.A., Socransky, S.S., 1981. *Wolinella* gen. nov., *Wolinella succinogenes* (*Vibrio succinogenes* Wolin *et al.*) comb. nov., and description of *Bacteroides gracilis* sp. nov., *Wolinella recta* sp. nov., *Campylobacter concisus* sp. nov., and *Eikenella corrodens* from humans with periodontal disease. Int. J. Syst. Bacteriol. 31, 432–435.

Tanner, A.C.R., Listgarten, M.A., Ebersole, J.L., 1984. *Wolinella curva sp. Nov., 'Vibrio succinogenes'* of human origin. Int. J. Syst. Bacteriol. 34, 275–282.

Thompson, L.M., Smibert, R.M., Johnson, J.L., Krieg, N.R., 1988. Phylogenetic study of the genus *Campylobacter*. Int. J. Syst. Bacteriol. 38, 190–200.

Totten, P.A., Fennell, C.L., Tenover, F.C., Wezenberg, J.M., Perine, P.L., Stamm, W.E., Holmes, K.K., 1985. *Campylobacter cinaedi* (sp. nov.) and *Campylobacter fennelliae* (sp. nov.): two new *Campylobacter* species associated with enteric disease in homosexual men. J. Infect. Dis. 151, 131–139.

Tu, Z.C., Eisner, B.N., Kreiswirth, Blaser, M.J., 2005. Genetic divergence of *Campylobacter fetus* strains of mammal and reptile origins. J. Clin. Microbiol. 43, 3334–3340.

Tyrrell, K.L., Citron, D.M., Warren, Y.A., Nachnani, S., Goldstein, E.J., 2003. Anaerobic bacteria cultured from the tongue dorsum of subjects with oral malodor. Anaerobe 9, 243–246.

Vandamme, P., De Ley, J., 1991. Proposal for a new family, Campylobacteraceae. Int. J. Syst. Bacteriol. 41, 451–455.

Vandamme, P., On, S.L., 2001. Recommendations of the subcommittee on the taxonomy of Campylobacter and related bacteria. Int. J. Syst. Evol. Microbiol. 51, 719–721.

Vandamme, P., Falsen, E., Pot, B., Hoste, B., Kersters, K., DeLey, J., 1989. Identification of EF Group 22 Campylobacters from gastroenteritis cases as *Campylobacter concisus*. J. Clin. Microbiol. 27, 1775–1781.

Vandamme, P., Falsen, E., Rossau, R., Hoste, B., Segers, P., Tytgat, R., DeLey, J., 1991. Revision of *Campylobacter*, *Helicobacter*, and *Wolinella* taxonomy: emendation of generic descriptions and proposal of *Arcobacter* gen. nov. Int. J. Syst. Bacteriol. 41, 88–103.

Vandamme, P., Vancanneyt, M., Pot, B., Mels, L., Hoste, B., et al., 1992. Polyphasic taxonomic study of the emended genus *Arcobacter* with *Arcobacter butzleri* comb. nov. and *Arcobacter skirrowii* sp. nov., an aerotolerant bacterium isolated from veterinary specimens. Int. J. Syst. Bacteriol. 42, 344–356.

Vandamme, P., Daneshvar, M.I., Dewhirst, F.E., Paster, B.J., Kersters, K., Goossens, H., Moss, C.W., 1995. Chemotaxonomic analyses of *Bacteroides gracilis* and *Bacteroides ureolyticus* and reclassification of *B. gracilis* as *Campylobacter gracilis* comb. nov. Int. J. Syst. Bacteriol. 45, 145–152.

Vandamme, P., Dewhirst, F.E., Paster, B.J., On, S.L. W., 2005. Genus I. *Campylobacter* Sebald and Veéron 1963, 907AL emend. Vandamme, Falsen, Rossau, Hoste, Segers, Tytgat and De Ley 1991a, 98. In: Brenner, D.J., Krieg, N.R., Staley, J.T., Garrity, G.M., (Eds.), Bergey's Manual of Systematic Bacteriology, vol. 2. Springer, New York. pp. 1147–1160.

Vandamme, P., Debruyne, L., De Brandt, E., Falsen, E., 2010. Reclassification of *Bacteroides ureolyticus* as *Campylobacter ureolyticus* comb. nov., and emended description of the genus *Campylobacter*. Int. J. Syst. Evol. Microbiol. 60, 2016–2022.

Veron, M., Chatelain, R., 1973. Taxonomic study of the genus *Campylobacter* Sebald and Veron and designation of the neotype strain for the type species, *Campylobacter fetus* (Smith and Taylor) Sebald and Veron. Int. J. Syst. Bacteriol. 23, 122–134.

Vinzent, R., Dumas, J., Picard, N., 1947. Septicemie grave au cours de la grossesse, due a un vibrion. Avortement consecutif. Bull. Acad. Nat. Med. 131, 90, (in French).

Zanoni, R.G., Debruyne, L., Rossi, M., Revez, J., Vandamme, P., 2009. *Campylobacter cuniculorum* sp. nov., from rabbits. Int. J. Syst. Evol. Microbiol. 59, 1666–1671.

Isolation, identification, and typing of *Campylobacter* strains from food samples

4

Omar A. Oyarzabal*, Catherine D. Carrillo**

*University of Vermont Extension, St. Albans, VT, United States;
**Canadian Food Inspection Agency, Ottawa, ON, Canada

4.1 INTRODUCTION

Campylobacteriosis is one of the most important foodborne bacterial diseases worldwide. *Campylobacter jejuni* and *C. coli* are the species most commonly associated with disease, and are responsible for more than 80% of the human cases worldwide, although recent work suggests that, as we improve surveillance in developing countries, more non-*C. jejuni/coli* infections will be identified (Platts-Mills et al., 2014). In developed countries, however, infections with *C. jejuni* and, to a lesser degree, *C. coli*, continue to be one of the most significant public health challenges.

The genus *Campylobacter* is comprised of about 30 taxa, of which 17 species are relevant to public health (Kaakoush et al., 2015). There is considerable variability in the annual incidence of campylobacteriosis by country (Table 4.1), and several epidemiological aspects of this disease are still not clearly understood. Campylobacteriosis has been primarily associated with handling or consumption of undercooked raw poultry and meat, the consumption of raw milk, raw shellfish (mainly oysters), and contaminated water (Greig and Ravel, 2009; Lévesque et al., 2013). Chicken meat, particularly, has a high prevalence of *C. jejuni/C. coli* and, in some cases, the number of contaminating cells is also high (Sproston et al., 2014). Thus, there is an increased chance of potential cross-contamination from raw to cooked meat, or from raw meat to other foods, when handling raw chicken in home kitchens (Luber, 2009; Luber et al., 2006). The underestimation of hygiene practices in kitchens to prevent cross-contamination is still very common, even among individuals that have suffered from campylobacteriosis (Millman et al., 2014).

The current methodology for isolation and identification focuses on clinical samples—mainly stool samples; food samples—primarily raw chicken meat and raw milk; and water samples. This chapter will review the methodology used to (1) isolate *C. jejuni* and *C. coli* from food samples, (2) identify isolates to the species level, and (3) type the isolates for epidemiological purposes. Because there are a large number of methods and assays developed for the isolation, identification, and

Table 4.1 Annual Incidence of Human Campylobacteriosis by Country (OzFoodNet Working Group, 2012; Crim et al., 2015; Kaakoush et al., 2015; Sears et al., 2011)

Country	Annual Incidence (Cases per 100,000 Population)
Australia	110
Denmark	42
German	53–80
Israel	90
Japan	1500
Canada	38
Guatemala	>180
New Zealand	160
Norway	30
The Netherlands	52
USA	14

typing of *Campylobacter* isolates from food samples, the Chapter will focus primarily on the methods that have been employed more extensively, and will provide an overview of some practical considerations for the adoption of these methods in food microbiology laboratories.

4.2 CHARACTERISTICS OF THE GENUS *CAMPYLOBACTER*

Table 4.2 shows some of the unique features of the genus *Campylobacter*. Some of these features have been exploited in the different methodologies designed to isolate *C. jejuni* and *C. coli* from clinical and food samples. Two important features that can benefit isolation are cultivation under microaerobic environments, and at 42°C. The term "microaerobic" refers to oxygen (O_2)concentration between 3 and 15%. The typical atmosphere for isolation of *C. jejuni*/*C. coli* contains around 5% O_2, 10% carbon dioxide (CO_2), and 85% nitrogen (N_2), which is an inert gas, although a concentration of 10% O_2 will be sufficient to grow most *Campylobacter* strains (Bolton and Coates, 1983). *C. jejuni* and *C. coli* do not grow well in media incubated under atmospheric conditions (20.94% O_2) during the initial isolation procedure. However, there are several reports of *C. jejuni* adapting to grow under aerobic conditions soon after initial isolation from clinical or food samples (Chynoweth et al., 1998; Rodrigues et al., 2015). In general, all *Campylobacter* species need higher concentrations of CO_2 (c. 10%) than the amount in ambient air, which is approximately 0.04%, or 400 parts per million. The term "capnophile" is employed to refer to microorganisms that require high concentrations of CO_2 to thrive and multiply.

Table 4.2 Main Characteristics of the Members of the Genus *Campylobacter* (Lastovica et al., 2014; On, 2001; Vandamme et al., 2005)

Feature	Values/Comments
Capnophilic	Some species require 35% CO_2 to grow
Catalase activity	Positive
Chemoorganotrophs	Do not ferment or oxidize carbohydrates
Energy	Obtained from amino acids or intermediates of the tricarboxylic acid cycle
High temperature for growth	42°C in case of thermotolerant species: *C. jejuni*, *C. coli*, *C. hyointestinalis*, *C. lari*, and *C. upsaliensis*[a]
Microaerobic atmosphere	O_2 concentration between 3 and 15%. Concentrations of 5% are commonly used for isolation
Minimal growth temperature	30°C
Mol.% G + C content of the DNA	27–31
Motility	Corkscrew-like darting motility observed with phase contrast or darkfield microscopy. High motility in fresh cultures
Shape	Spiral, S-shaped, or gull-winged-shaped when two cells form short chains. Cells in old cultures can form spherical or coccoid bodies
Special requirements to grow	Some species require hydrogen or formate with fumarate (electron donors) to grow in microaerobic conditions. If not, anaerobiosis becomes an optimal growth condition for these species

[a]*Other species can also grow at this temperature, but 42°C has been extensively used for isolation of* C. jejuni *and* C. coli *from food and clinical samples.*

The microbiology literature has a large body of information on how to provide for aerobic conditions during the culturing of microorganisms. We currently know the importance of the diffusion of O_2 into culture media for different metabolic activities, such as electron transport, etc. (Somerville and Proctor, 2013). For the cultivation of aerobic bacteria, agitation of the media during culturing, and optimal flask-to-medium ratio provides for an appropriate diffusion of O_2 through culture broths. Now we also know that liquid media naturally create a microaerobic area between the surfaces of the liquid and the deeper areas in the liquid, where O_2 has a limited availability for bacterial growth (Fenchel and Finlay, 2008; Somerville and Proctor, 2013). The current isolation procedures using liquid rely on the generation of microaerobic atmospheres during cultivation, based on the replacement of the atmosphere with a microaerobic gas mix containing approximately 5% O_2, or by the sequestration of O_2 to reduce it to approximately 5%. Yet, the enrichment of samples in liquid media may provide enough of a microaerobic type of condition for *Campylobacter* spp. to grow to detectable levels (Zhou et al., 2011; Oyarzabal et al., 2013).

C. jejuni has several mechanisms to cope with O_2 species that have the potential to destroy bacterial cells, such as hydrogen peroxide (Atack et al., 2008). In general, the more complex the media are when isolating *Campylobacter* spp., the less O_2 will be present for the growth of *C. jejuni*. *C. jejuni* can therefore take advantage of microniches with lower O_2 concentration by detecting these microniches through oxygen-sensing mechanisms, and moving toward them by using their active flagella. These properties make *C. jejuni* a very good survivor in different environmental conditions, especially those conditions found in foods of poultry and meat origins, particularly at lower temperatures of 4–5°C that are the normal temperatures for storage of meat food products (Chynoweth et al., 1998; Garénaux et al., 2008). In addition, commensalism with other bacteria increases the tolerance to O_2 by *C. jejuni* (Reuter et al., 2010). Recently, the study from Hilbert et al. (2010) has demonstrated that *C. jejuni* can survive aerobic conditions for at least 48 h, when cultivated in association with *Pseudomonas putida*. Although there was large variability in the length of time that different *C. jejuni* survived coincubation with *P. putida*, all isolates were viable for longer than 18 h than in single culture studies. An interesting finding from this study was the variability in the survival rate of the *C. jejuni* isolates, with the isolates from broiler fecal samples having a significantly reduced survival rate, in comparison with the isolates collected from chicken meat or human samples. These findings may suggest that, as *C. jejuni* strains leave the intestinal tract of domestic animals, they are exposed to different environmental conditions and become more resistant to different stress situations, including higher O_2 concentrations (Hilbert et al., 2010; Oh et al., 2015).

The way *C. jejuni* and *C. coli* obtain energy is different from other bacterial pathogens, such as *Salmonella* and *Listeria*. The genes that encode pathways for utilization of carbohydrates are either incomplete or entirely missing in *C. jejuni* (Hofreuter, 2014). The main sources of energy are amino acids, or intermediates of the tricarboxylic acid cycle, such as acetate, aspartate, glutamate, and pyruvate (Table 4.2). Aspartate is a precursor for the biosynthesis of lysine, methionine, threonine, and isoleucine, and together with glutamate, proline, and serine, aspartate is considered an amino acid that promotes growth in most of the *C. jejuni* strains (Guccione et al., 2008; Hofreuter et al., 2008, 2014). In *C. jejuni*, aspartate is deaminated to fumarate, a reaction that helps *C. jejuni* cope with its recovery from cell injury. In turn, fumarate can be converted to oxaloacetate in order to generate a substrate for the gluconeogenesis and synthesis of essential carbohydrates (Bieche et al., 2012; Guccione et al., 2008; Sellars et al., 2002). Propionic acid, however, appears to be utilized only by *C. coli*, because *C. jejuni* does not appear to have the genes needed for propanoate metabolism. Thus, the utilization of propionic acid may be a unique feature that could be used to differentiate between *C. coli* and *C. jejuni* (Occhialini et al., 1996; Wagley et al., 2014).

C. jejuni has several mechanisms that enable survival in the environment. One of the survival mechanisms for this pathogen is its ability to resist digestion, and survive inside protozoa, such as amoeba and *Tetrahymena pyriformis*. This particular survival mechanism has been associated with the survival in unfavorable environment, and has been linked to the resistance of these bacteria to antimicrobials substances (Snelling et al., 2005). A particular characteristic of *C. jejuni* is that this bacterium can express virulence and produce disease in humans, while the same strain establishes a commensal relationship in chickens and other hosts. Mechanisms enabling growth and survival in different species are complex, and not well-understood; however, it appears that isolates most commonly associated with human infection are those that are adapted to grow well in different hosts (Dearlove et al., 2015).

Despite its fastidious nature, and relatively slow growth rate, *C. jejuni* competes very well with the microbiota in the human intestine to then adhere to, and invade, epithelial cells. One of the attributes that may influence the success of this organism in the gut environment is that Campylobacters are typically highly motile, with a characteristic corkscrew-like movement driven by means of a single polar unsheathed flagellum, at one or both ends of the cell. This particular type of motility is an important feature aiding in the identification of *Campylobacter* spp. using phase–contrast microscopy, following isolation from clinical or food samples. In addition, several adhesion factors (called adhesins) enabling host epithelial cell invasion have been well-characterized (Ó Cróinín and Backert, 2012). The genes responsible for the production of these proteins are good targets for PCR assays for identification of *C. jejuni*.

4.3 ISOLATION

Some of the key attributes that have been highlighted in the previous paragraphs have been incorporated in different protocols for the isolation of *Campylobacter* spp. There is no gold standard for the routine isolation of all *Campylobacter* species. All the "reference" methods are based on direct plating, on selective plates, and/or coupled with the enrichment of the samples and then transfer to selective plates (ISO, 2006; Food Safety and Inspection Service, 2014; Carrillo and Iugovaz, 2014). Only direct plating is used in the USA for the detection of *Campylobacter* spp. in processed poultry carcasses (Food Safety and Inspection Service, 2014). However, the application of a microaerobic atmosphere with a filtration method on selective agar plates, with the selectivity provided by cefoperazone and amphotericin B, is one of the most simple and practical combination for the isolation of *Campylobacter* spp. from foods. Some of the methods that generate microaerobic atmosphere also generate some hydrogen, which was shown to be important for isolation of *C. upsaliensis* (Goossens et al., 1990). However, the priority species, *C. jejuni* and *C. coli*, can grow well in microaerobic atmospheres on selective media, without the presence of hydrogen. A simplified methodology for isolation is presented in Fig. 4.1. The methodology of isolation starts with the collection of food (e.g., poultry meat sample) and the enrichment of the food sample in a selective enrichment. The enrichment broth does not need to be rich, and the ratio of meat:enrichment appears to vary. If the meat sample is incubated with the enrichment, a ratio of 1:4 appears to be the minimal ratio to employ (Oyarzabal and Liu, 2010). However, if the meat is removed after a short rinse in the enrichment broth, and the enrichment broth is incubated without the meat sample, a ratio of 1:2 can be used for isolation purposes (Oyarzabal et al., 2013).

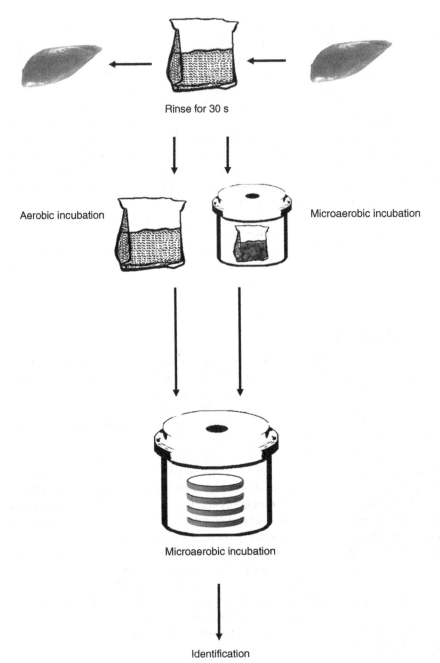

Rinse for 30 s

Aerobic incubation   Microaerobic incubation

Microaerobic incubation

Identification

FIGURE 4.1 Method Describing the Most Practical Method to Isolate *Campylobacter* spp. from Chicken Meat

The baseline survey of processed chicken carcasses done in the USA and Canada used a protocol where 30 mL of a carcass rinse were mixed with 30 mL of double strength Bolton broth (Food Safety and Inspection Service, 2009). However, the current methodology used in the USA by the Food Safety and Inspection Service of the US Department of Agriculture does not include an enrichment step anymore, and only the results from direct plating are recorded (Food Safety and Inspection Service, 2013a, 2014).

The addition of selective antibiotic agents to the enrichment broth appears to be more important for the initial isolation because the enrichment time is long, usually 24–48 h. Therefore, inhibiting competing microflora is more important in the first 24 h of isolation for the successful detection of *Campylobacter* in the sample (Gharst et al., 2013). Cefoperazone (sodium salt) at concentrations of 33 mg/L of medium, and amphotericin B at concentration of approximately 4–10 mg/L of medium is a combination that inhibits competing microflora, while apparently not exerting too much inhibition of *Campylobacter* strains. If the transfer of the enrichment broth to plate media will be done using a filter membrane, there is no need for the addition of other antimicrobials, such as vancomycin (usually added at concentrations of 20 mg/L of medium). We also have anecdotal evidence that vancomycin may actually hinder the isolation of some *Campylobacter* strains; therefore, we prefer the use of filter membranes to provide for a methodology that prevents competing microflora but has minimal inhibitory impact on the growth of *Campylobacter* strains (Gharst et al., 2013; Speegle et al., 2009).

Some of the most commonly used enrichment broths are described in Table 4.3. Buffered peptone water is a simple enrichment broth that can be successfully incorporated for isolation purposes (Oyarzabal et al., 2007, 2013), especially the rinsing of the meat sample and the incubation of buffered peptone water (enrichment medium) without the meat sample (Figure 4.1), and without the addition of blood (Liu et al., 2009). The incubation of enrichment broths have always been done under atmospheres containing a reduced O_2 level, usually the traditional "microaerobic" atmosphere comprised of 5% O_2, 10% CO_2, and 85% N_2. But some of our research shows that enrichment broths naturally generate microaerobic environment during static incubation, and therefore there is no need to add a microaerobic atmosphere for the isolation of *Campylobacter* from food samples (Oyarzabal et al., 2013; Zhou et al., 2011).

The time for enrichment of the samples can vary. Most of the semiautomated, commercial PCR systems, such as the BAX (Dupont, Qualicon, Wilmington, DE, USA) and iQ-Check (BioRad Laboratories, Hercules, CA, USA) require an enrichment of approximately 24–48 h. However, these systems have been validated only with poultry carcass rinses that may have a higher contamination rate, and probably more cells of *Campylobacter* per unit than retail poultry meat, at least in North America. In a recent study on retail poultry meat, we found only a minimal benefit to increasing enrichment to 48 h, with a 1.9% increase in positive samples, (Carrillo et al., 2014), but, in general, when isolating *Campylobacter* spp. from poultry meat, the recommended time is still 48 h. After incubation, a portion of the enrichment broth is transferred onto agar

Table 4.3 Enrichment Broths and Pate Media Commonly Used for Isolation of *Campylobacter* spp.

Category	Medium	References
Enrichment	Bolton broth: Enzymatic digest of animal tissue, lactalbumin hydrolysate, yeast extract, sodium chloride, hemin, sodium pyruvate, α-ketoglutamic acid, sodium metabisulfite, sodium carbonate[a]	Baylis et al. (2000); Bolton and Coates (1983)
	Buffered peptone water: Peptone, sodium chloride, disodium phosphate, monopotassium phosphate[a]	Gharst et al. (2013); Oyarzabal et al. (2007); Oyarzabal et al. (2013)
	Preston broth: Lab-Lemco powder, peptone, sodium chloride[a,b]	Bolton and Robertson (1982)
Plate	Charcoal-based: charcoal cefoperazone deoxycholate agar (CCDA)[a]	Hutchinson and Bolton (1984); Oyarzabal et al. (2005)
	Blood-based: Brucella agar, Columbia agar, etc.	Gharst et al. (2013)

Other formulations are available. Some formulations may include the supplementation with sodium pyruvate, sodium metabisulfite, and ferrous sulfate.
[a]*Basal medium can be purchased from Acumedia (Lansing, MI), Oxoid (Thermo-Fisher), etc.*
[b]*Formulations from Oxoid (Thermo-Fisher).*

plates. The basal medium for agar plates has traditionally incorporated blood or charcoal to help reduce the O_2 in the atmosphere, and facilitate the growth of *Campylobacter* spp. (Bolton and Coates, 1983; Bolton and Robertson, 1982; Hutchinson and Bolton, 1984). Blood was the substance originally added to agar plates for the isolation of *Campylobacter* until charcoal was found to be a suitable replacement (Bolton and Coates, 1983). Activated charcoal has been used since the beginning of the 1900s for the adsorption of gases, vapors, and colloidal solids (Pan and Van Staden, 1998). In microbiology media, charcoal inactivates hydrogen peroxide and helps quench O_2. While presently there are many systems to generate a microaerobic atmosphere to promote the growth of *Campylobacter* spp., the addition of charcoal or blood is still important to create differential media with a background that improves the visualization and recognition of *Campylobacter* colonies. As a matter of fact, *Campylobacter* will grow well on agar media without blood or charcoal, if all the other growth conditions are met; however, the colonies will be almost transparent, and very difficult to visualize or differentiate from other bacterial colonies.

Currently, charcoal-cefoperazone-deoxycholate agar (CCDA) is the most broadly used agar plate for isolation of Campylobacters, and it is likely that blood agar plates will be replaced by other differential plates in the future, due to increasing cost of animal blood. We have found that the same antibiotic combination and amount described for enrichment broth (cefoperazone salt: 33 mg; amphotericin B: 4–10 mg; both per liter of medium) is suitable for plating media used in the isolation of *Campylobacter* from food samples. There has been a long discussion, for more than five decades, on the best antibiotic supplementation for the isolation of *Campylobacter* spp. Some scientists have pursued the search for an "optimal" combination, and have incorporated antibiotics developed for the isolation of other enteric bacteria (Corry et al., 1995); others have added extensive antibiotics (Line, 2001; Line et al., 2008); and, more recently, others have added new antibiotics to traditional plates in the hope of increasing selectivity (Chon et al., 2012), especially against *Escherichia coli* strains expressing extended-spectrum beta-lactamase (Smith et al., 2015). Yet, we advocate for less use of antibiotics and the incorporation of mechanical means, such as filter membranes, to separate *Campylobacter* from the rest of the microflora present in the food sample. This approach is also applicable to the isolation of *Campylobacter* spp. from any other samples, such as clinical, veterinary, and environmental and water samples (Gharst et al., 2013; Carrillo et al., 2016).

Several new chromogenic agars have been developed in recent years for the isolation of *C. jejuni* and *C. coli*. The chromogenic agars currently on the market in the USA, Europe, and some countries in Latin America, have been validated for beef, pork, poultry meat, carcass rinse, and environment samples. The available chromogenic agars include CampyFood Agar and CASA (*Campylobacter* Selective Agar) by bioMérieux, RAPID'*Campylobacter*/Agar by Bio-Rad Laboratories, and Brilliance CampyCount Agar by Oxoid Ltd (part of Thermo Fisher Scientific). Most of these chromogenic agars perform similarly to other traditional plates, such as CCDA; yet, some of these chromogenic agar plates may not provide an ideal selectivity either, and other bacterial species may grow on the plate. Some may not even be completely differential (Ahmed et al., 2012; Habib et al., 2008, 2011; Seliwiorstow et al., 2014; Teramura et al., 2015).

The isolation of *Campylobacter* from foods relies also on an incubation temperature of 42°C. Growth at this temperature is not unique to *Campylobacter*, but it is not common for most of the bacteria present in food samples, except some enteric organisms such as *E. coli*. This high temperature helps reduce the competition from other bacteria and yeasts, and should be kept throughout the isolation procedure (enrichment and growth on selective agar plates). The only time when this temperature is not advisable is when trying to isolate a nonthermotolerant species, or species other than *C. jejuni*, *C. coli*, *C. lari*, or *C. upsaliensis*. In these cases, the recommended temperature for isolation should be 37°C. Although some enrichment protocols haven an initial temperature of 37°C for the first 3–4 h of enrichment, to help recovered injured *C. jejuni/C. coli* cells, there is a lack of strong scientific evidence to suggest that temperature alone will be enough for recovery. A protocol that really targets injured cells should include other variable, such as no antibiotic addition for the first 6–8 h, a fact that will make the protocol very cumbersome, and difficult to incorporate by food microbiology laboratories.

4.4 IDENTIFICATION OF SPECIES

Some phenotypic tests have been designed for the identification of species, but these tests have not found a practical application in laboratories because of their lack of reliability, and the fact that they are time-consuming tests. It is important to remember that most strains of *Campylobacter* do not use carbohydrates to generate energy and, therefore, traditional phenotypic tests based on carbohydrate utilization are not conducive to species identification (Lastovica et al., 2014; On, 2001). Given the frequency of isolation of *C. jejuni* and *C. coli*, reliable identification of these species is critical. Traditional biochemical approaches for distinguishing *C. coli* from *C. jejuni* relied on results of the hippurate hydrolysis test. Using this test, *C. jejuni* usually produces a positive reaction, while *C. coli* produces a negative reaction. However, some strains of *C. jejuni* subsp. *jejuni* can also produce negative reactions. In the past 20 years, polymerase chain reaction (PCR) assays targeting the hippuricase gene have been more successful in differentiating these two species because although *C. jejuni* carries the gene, in some strains it may not be active, while all *C. coli* strains appear to be missing this gene. Another distinguishing feature of *C. coli* is the ability to utilize propionate as a sole carbon source, while *C. jejuni* cannot (Occhialini et al., 1996; Wagley et al., 2014).

There are only a few tests that are available for the identification of *Campylobacter* spp., or *C. jejuni/C. coli/C. lari*. The tests that have been validated for raw chicken meat, chicken carcass rinse, or raw milk include two lateral flow devices: (1) Singlepath *Campylobacter* (Merck KGaA) for raw ground chicken, raw ground turkey, and pasteurized milk; and (2) Veriflow *Campylobacter* (Invisible Sentinel, Inc.) for chicken carcass rinse; one ELISA test, VIDAS *Campylobacter* by bioMérieux for meat products, environment samples, raw pork, raw chicken breast, processed chicken nuggets, chicken carcass rinse, and turkey carcass sampled with sponge; and the two semiautomated PCR assay already mentioned (BAX System and iQ-Check *Campylobacter* Real-Time PCR). The latex agglutination test that is most commonly used for identification is Microgen *Campylobacter* (Microgen Bioproduct Ltd., Surrey, UK, available in the USA through several distributors, such as Hardy Diagnostics). Most latex agglutination tests target antigenic elements that are similar in C. *jejuni*, *C. coli*, and *C. lari*. The US Department of Agriculture, through the Microbiology Laboratory Guidebook, also accepts phase-contrast microscopy as a means to confirm presumptive colonies on plates (Food Safety and Inspection Service, 2013b). The Health Canada method recommends phase–contrast microscopy as a means of presumptive identification of *Campylobacter* spp., but requires further confirmatory tests (Carrillo and Iugovaz, 2014)

The most reliable methods for identification of species are PCR assays. These assays have been multiplexed and incorporated into the few real-time PCR assays that are commercially available. These PCR assays target DNA segments from genes, or DNA sequences that are unique to each species. Table 4.4 summarizes some of the pathogenicity genes that have been used, or are available, for the identification of *Campylobacter* spp. through PCR assays. The most commonly used protocol is the multiplexing of primers where there are at least four primers to identify *C. jejuni* and *C. coli*. Most of the food isolates can be confirmed with a multiplex PCR targeting *C. jejuni/C. coli*, as these two species are the predominant species found in food samples. A summary of the primers, and their DNA targets, that have been validated

Table 4.4 Major Adhesin Proteins Described for *C. jejuni* and *C. coli* Segments of the Sequences of the Genes Encoding for Some of These Adhesins Have Been Incorporated Into PCR Assays for Species Identification

Species	Target	Protein Size	References
C. jejuni	*Campylobacter* adhesion to fibronectin—CadF	37-kDa outer membrane protein	Konkel et al. (1997); Nayak et al. (2005)
	Fibronectin like protein A—FlpA	46-kDa protein	Konkel et al. (2010); Eucker and Konkel (2012)
	Lipoprotein A—JlpA	43-kDa protein	Jin et al. (2003)
	Major outer membrane protein—MOMP	59-kDa protein	Moser et al. (1997)
	Protein 95—p95	95-kDa protein	Kelle et al. (1998)
C. coli	Lipoprotein—CeuE	37-kDa lipoprotein	Nayak et al. (2005)

Modified from Oyarzabal and Backert (2016).

with large numbers of *C. jejuni/C. coli* strains is presented in Table 4.5. This table is by no means comprehensive, but provides the most commonly used primer sets for singleplex or multiplex identification of *C. jejuni/C. coli* species.

4.5 TYPING METHODS

After the isolation of *Campylobacter* strains, and the identification of the strains to the species level, it may be useful to further characterize the isolates. One of the most important epidemiological questions when dealing with cases of human campylobacteriosis is to identify the sources of these strains. For instance, in the analysis of a cluster of cases of human campylobacteriosis, a food may be suspected as the source of contamination. But just the simple isolation of *Campylobacter* spp. from the implicated food is not enough to make a definitive link between the food and the human cases. The next step is to characterize the strains with a typing method to see if the food strains "match" with the human strains. In this section of the Chapter we review the typing methods most commonly used in surveillance programs worldwide. For a more comprehensive review of the typing methods for *C. jejuni/C. coli*, please refer to Carrillo and Oyarzabal (2014).

The first two typing schemes for *Campylobacter* species were developed in the 1980s, and were based on serotyping. One serotyping scheme recognizes heat-stable or somatic O-antigens, and was described by Penner and Hennessy (1980). The major antigenic components in this scheme are thought to be lipopolysaccharides or lipooligosaccharides. The second scheme is based on heat-labile antigens, and was proposed by Lior et al. (1981). The flagellar proteins have been considered the

Table 4.5 Most Commonly Used Primers to Identify *C. jejuni* and *C. coli*. Most of These Primers Have Been Used in Multiplex PCR Assays

Species Identified	Target Gene	Primer Name	Sequences (5'–3')	Product Size (bp)	References
C. coli and *C. jejuni*	CadF gene (adhesin, fibronectin)	CadF2B CadR1B	TTGAAGGTAATTTAGATATG CTAATACCTAAAGTTGAAAC	400	Konkel et al. (1999); Cloak and Fratamico (2002); Oyarzabal et al. (2005, 2008)
C. coli	Aspartokinase gene	CC18F CC519R	GGTATGATTTCTACAAAGCGAG ATAAAAGACTATCGTCGCGTG	500	Linton et al. (1997); Potturi-Venkata et al. (2007); Miller et al. (2010); Carrillo et al. (2016)
	ceuE gene (siderophore transport)	COLI COL2	ATGAAAAAATATTTAGTTTTTGCA ATTTTATTATTTGTAGCAGCG	894	Gonzalez et al. (1997); Cloak and Fratamico (2002); Oyarzabal et al. (2005, 2008)
	Aspartokinase gene	ask-F-JK ask-R-JK	GGCTCCTTTAATGGCCGCAAGATT AGACTATCGTCGCGTGATTAGCG	306	Zhou et al. (2011)
C. jejuni	Hippuricase gene	HipO-F HipO-R	GACTTCGTGCAGATATGGATGCTT GCTATAACTATCCGAAGAAGCCATCA	344	Persson and Olsen (2005); Potturi-Venkata et al. (2007); Carrillo et al. (2016)
	glyA genen (glycine)	glyA-F-JK glyA-R-JK	TGGCGGACATTTAACTCATGGTGC CCTGCCACAACAAGACCTGCAATA	264	Zhou et al. (2011)

Table 4.6 Summary of Most Common Typing Methods Used to Characterize *C. jejuni/C. coli* Strains

Method	Advantages	Disadvantages
Pulsed field gel electrophoresis (PFGE)	• Standardized protocols enable a 24 h turnaround time for results • High resolution supporting outbreak analysis • Large set of results available in surveillance databases internationally	• Requires specialized equipment • Need for special software to interpret the results of a large number of strains • Some strains are untypeable
Multilocus Sequence Typing (MLST) and extended MLST (eMLST)	• Easy data sharing through central repositories (e.g., PubMLST website, http://pubmlst.org) • Most common typing method used internationally for epidemiological characterization, and source tracking of many infectious pathogens • Data are easy to compare among laboratories	• Labor-intensive if not automated • Some strains are difficult to type due to difficulty in amplifying targets • Standard MLST schemes have limitations for shorter-term epidemiological studies, and routine surveillance
Whole Genome Sequencing (WGS)	• Highest resolution, highest discriminatory power • Typing data for most DNA-based methods can be derived in silico • Phenotypic properties can also be derived from WGS data (e.g., AMR, virulence genes) • Enables determination of relationship among strains • Can be applied with no a priori knowledge of bacterial species	• High cost • Requires access to specialized equipment • Extensive computational requirements
Comparative Genomic Fingerprinting (CGF)	• Low cost, high throughput • In most cases, strains cluster according to evolutionary relationship • Standardized protocols with limited to no variability in results • Over 25,000 strains typed. Highly portable	• Not implemented internationally • Databases are not yet publicly available

Modified from Carrillo and Oyarzabal (2014).

principal antigenic element targeted in this heat-labile scheme. However, in a study where strains were tested with the heat-labile scheme and a PCR-based restriction fragment length polymorphism (RFLP) targeting the *fla*A and *fla*B genes, there was little conservation of flagellar RFLP patterns within serotyping groups. These findings suggest a large flagellar heterogeneity within serogroups (Mohran et al., 1996). Neither of these serotyping schemes is commonly used today, and few laboratories have stocks of antisera to perform these assays.

Other typing assays developed for *Campylobacter* spp. include plasmid typing and PCR-based typing schemes. Plasmids have not provided a suitable typing method for *C. jejuni* and *C. coli* due to the large variability in the prevalence (19–95%), the number (1–5), and the sizes of the plasmids (2–208 kbp) found in different strains (Austen and Trust, 1980; Lee et al., 1994; Tenover et al., 1985). The typing schemes based on PCR assays (e.g., amplification of repetitive sequences (REP-PCR), amplification of variable-number tandem repeat sequences, *fla*A-RFLP, amplified fragment length polymorphism, etc.) have been used in several research studies, and some are of low cost. Yet, most of the PCR methods are of low discriminatory power, and very few have been automated (e.g., REP-PCR). In addition, some of these PCR methods are lacking a standardized protocol to ensure repeatability (Carrillo and Oyarzabal, 2014).

One method that was discovered in the 1980s, pulsed-field gel electrophoresis (PFGE), has become an important typing tool with application in surveillance systems worldwide. PFGE is a typing method in which the whole genome of an organism is cut using rare-cutting restriction enzymes, and then the DNA fragments are separated on the basis of their sizes. The separation of large DNA segments, the most important discovery in PFGE, is achieved by an abrupt change in the electric field during the electrophoretic run to make the DNA molecules change directions. Thus, DNA molecules are forced to undergo size-dependent reorientation before they can move through the gel. PFGE can resolve DNA molecules up to 12 Mb in size, and has become an important molecular method for typing foodborne bacterial pathogens, especially *C. jejuni* and *C. coli* (Ray and Schwartz, 2014). Although some reports suggest that the PFGE profiles of some strains may change over time due to phage infections or other genomic rearrangements (Barton et al., 2007), PFGE is suitable for establishing short-term epidemiological relationships of *C. jejuni*/*C. coli* strains. PFGE enables rapid sharing of results for fast and efficient identification, investigation, and response to *Campylobacter*-associated outbreaks. There are several standardized protocols to perform PFGE on *Campylobacter* isolates (Center for Disease Control and Prevention, 2016; Ribot et al., 2001; Zhou and Oyarzabal, 2015), and PFGE is one of the main typing methods used by PulseNet, the national molecular typing network (Swaminathan et al., 2001). The large variability in the PFGE profiles of *C. jeuni* and *C. coli* is a limitation for long-term studies of the movement of these strains in natural environments, and the grouping of isolates within these two species is not always consistent with PFGE. Therefore, PFGE should not be used to identify *Campylobacter* isolates to the species level.

For long-term epidemiological studies of *C. jeuni* and *C. coli*, multilocus sequence typing (MLST) has proven to be extremely valuable, and has been extensively deployed worldwide. This typing scheme is based on the amplification and sequencing of several housekeeping genes that are distributed throughout the genome of the organism. Each unique sequence profile is assigned an allele number, and the combination of the different alleles determines the sequence type (ST) for a given strain. Strains that have four or more similar alleles in comparison with a central genotype are assigned to the same clonal complex or lineage (Dingle et al., 2002; Maiden et al., 1998). The advantage of this method is that sequence information generated in different laboratories is readily comparable, making the method highly portable. One drawback of this approach is that several primer sets are necessary to account for the variability among *Campylobacter* isolates, and typing of unusual genotypes may be problematic due to difficulties in amplifying all of the target genes. Also, MLST is often not sufficiently discriminatory for short-term epidemiological studies, such as outbreak investigations. To address the need for increased discriminatory power, an extended MLST (eMLST) typing scheme that incorporates antigenic gene targets, including short variable regions (SVR) of the *flaA* and *flaB* genes and the *porA* gene, was developed (Dingle et al., 2008; Carrillo et al., 2012). Table 4.6 shows some of the advantages and disadvantages of the most common typing methods used for characterization of *C. jejuni/C. coli* strains.

The introduction of rapid, low cost methods for whole-genome sequencing (WGS) has resulted in a significant shift in the methodology for typing bacterial foodborne strains. With increasingly widespread availability of instrumentation and computational infrastructure, costs for WGS of bacteria are now comparable to costs for traditional MLST analyses. This is particularly true for bacteria with small genomes, such as *Campylobacter* spp. that typically range in size between 1.6 and 2.0 million bases. Traditional MLST or other sequence-based typing data can be easily derived from WGS data (Carrillo et al., 2012; Colles and Maiden, 2012), but the availability of the whole genome enables the use of more discriminatory extended typing schemes. For example, the 53-gene ribosomal MLST (rMLST) is based on genes encoding ribosomal proteins that are universally present in all bacteria (Jolley et al., 2012). This scheme can distinguish isolates to the strain level (Maiden et al., 2013). To achieve higher levels of discrimination, core genome MLST (cgMLST) based on the analysis of hundreds of genes that are conserved among all strains within a species, or whole genome MLST (wgMLST) that incorporates all of the genes in a genome, can be applied (Cody et al., 2013; Kovanen et al., 2014; Winstantley et al., 2015). In cases where sufficient discriminatory power is not achieved with any of these methods, comparative genomic analyses can be used to determine single nucleotide polymorphisms (SNPs) or differences in the presence of accessory genes among isolates (Biggs et al., 2011; Kivistö et al., 2014; Revez et al., 2014).

While WGS analysis has been increasingly used in recent years for typing *Campylobacter* spp., for large scale studies of thousands or tens of thousands of *Campylobacter* isolates, there remains a need for rapid, automatable, and inexpensive methods. One such method, Comparative Genomic Fingerprinting (CGF) was developed based on the accessory genes identified through genomic analyses of hundreds of *Campylobacter* isolates (Taboada et al., 2004, 2012). Typing by CGF is based on assessing the presence of absence of 40 *Campylobacter* genes by multiplex PCR, using eight five-plex PCR reactions. Unlike PFGE, and sequencing approaches for *Campylobacter* typing, CGF does not require specialized equipment, and can be conducted in any lab with capacity for PCR analysis and traditional gel electrophoresis. CGF provides comparable discriminatory power to eMLST schemes (Carrillo et al., 2012; Clark et al., 2012), and has now been widely deployed in Canadian surveillance activities (Carrillo et al., 2012; Deckert et al., 2014).

The selection of the most appropriate method for typing *Campylobacter* spp. will be influenced by a number of factors, including the number of strains to be analyzed, the purpose of the study (e.g., short-term epidemiology versus long-term epidemiology), the need to compare results to public databases and, of course, funding available for the activity. WGS is the most comprehensive approach for *Campylobacter* typing, and data generated worldwide can be easily compared for both long-term and short-term epidemiological analysis. However, WGS is more expensive than methods such as PFGE and CGF, and may not be appropriate for large scale studies, or in cases where comparisons must be made to current or historical collections typed by different approaches (e.g., PFGE).

REFERENCES

Ahmed, R., Leon-Velarde, C.G., Odumeru, J.A., 2012. Evaluation of novel agars for the enumeration of *Campylobacter* spp. in poultry retail samples. J. Microbiol. Methods 28, 304–310.

Atack, J.M., Harvey, P., Jones, M.A., Kelly, D.J., 2008. The *Campylobacter jejuni* thiol peroxidases Tpx and Bcp both contribute to aerotolerance and peroxide-mediated stress resistance but have distinct substrate specificities. J. Bacteriol. 190, 5279–5290.

Austen, R.A., Trust, T.J., 1980. Detection of plasmids in the related group of the genus *Campylobacter*. FEMS Microbiol. Lett. 8, 201–204.

Barton, C., NG, L.-K., Tylor, S.D., Clark, C.G., 2007. Temperate bacteriophages affect pulsed-field gel electrophoresis patterns of *Campylobacter jejuni*. J. Clin. Microbiol. 45, 386–391.

Baylis, C.L., MacPhee, S., Martin, K.W., Humphrey, T.J., Betts, R.P., 2000. Comparison of three enrichment media for the isolation of *Campylobacter* spp. from foods. J. Appl. Microbiol. 89, 884–891.

Bieche, C., de Lamballerie, M., Chevret, D., Federighi, M., Tresse, O., 2012. Dynamic proteome changes in *Campylobacter jejuni* 81–176 after high pressure shock and subsequent recovery. J. Proteomics 75, 1144–1156.

Biggs, P.J., Fearnhead, P., Hotter, G., Mohan, V., Collins-Emerson, J., Kwan, E., Besser, T.E., Cookson, A., Carter, P.E., French, N.P., 2011. Whole-genome comparison of two *Campylobacter jejuni* isolates of the same sequence type reveals multiple loci of different ancestral lineage. PLoS One 6, e27121.

Bolton, F.J., Coates, D., 1983. Development of a blood-free *Campylobacter* medium: screening tests on basal media and supplements, and the ability of selected supplement to facilitate aerotolerance. J. Appl. Bacteriol. 54, 115–125.

Bolton, F.J., Robertson, L., 1982. A selective medium for isolating *Campylobacter jejuni/coli*. J. Clin. Pathol. 35, 462–467.

Carrillo C, Iugovaz I. 2014. Isolation of Campylobacter from food, MFLP-46. In: Compendium of analytical methods, vol. 3. Health Products and Food Branch, Health Canada, Ottawa, Ontario.

Carrillo, C.D., Oyarzabal, O.A., 2014. Molecular typing of *Campylobacter jejuni*. In: Oyarzabal, O.A., Kathariou, S. (Eds.), DNA methods in food safety: molecular typing of foodborne and waterborne bacterial pathogens. Wiley Blackwell, Chichester, UK.

Carrillo, C., Kruczkiewicz, P., Mutschall, S., Tudor, A., Clark, C., Taboada, E.N., 2012. A framework for assessing the concordance of molecular typing methods and the true strain phylogeny of *Campylobacter jejuni* and *C. coli* using draft genome sequence data. Front. Cell Infect. Microbiol. 2, 57.

Carrillo, C.D., Plante, D., Iugovaz, I., Kenwell, R., Belanger, G., Boucher, F., Pouline, N., Trottier, Y.-L., 2014. Method dependent variability in determination of prevalence of *Campylobacter jejuni* and *Campylobacter coli* in Canadian retail poultry. J. Food Protect. 77, 1682–1688.

Carrillo, C.D., Kenwell, R., Iugovaz, I., Oyarzabal, O.A., 2016. Recovery of *Campylobacter* spp. from food and environmental sources. In: Butcher, J., Stinzi, A. (Eds.), Methods in molecular biology. *Campylobacter jejuni*: methods and protocols. Humana Press, a part of Springer Science + Business Media, Totowa, NJ.

Center for Disease Control and Prevention, 2016. PulseNet Pathogens & Protocols. PulseNet. Available from: http://www.cdc.gov/pulsenet/pathogens/index.html

Chon, J.W., Hyeon, J.Y., Yim, J.H., Kim, J.H., Songm, K.Y., Seo, K.H., 2012. Improvement of modified charcoal–cefoperazone–deoxycholate agar by supplemented with a high concentration of polymyxin B for detection of *Campylobacter jejuni* and *C. coli in* chicken carcass rinses. Appl. Environ. Microbiol. 78, 1624–1626.

Chynoweth, R.W., Hudson, J.A., Thom, K., 1998. Aerobic growth and survival of *Campylobacter jejuni* in food and stream water. Lett. Appl. Microbiol. 27, 341–344.

Clark, C.G., Taboada, E., Grant, C.C.R., Blakeston, C., Pollari, F., Marshall, B., Rahn, K., Mackinnon, J., Daignault, D., Pillai, D., Ng, L.-K., 2012. Comparison of molecular typing methods useful for detecting clusters of *Campylobacter jejuni* and *C. coli* isolates through routine surveillance. J. Clin. Microbiol. 50, 798–809.

Cloak, O.M., Fratamico, P.M., 2002. A multiplex polymerase chain reaction for the differentiation of *Campylobacter jejuni* and *Campylobacter coli* from a swine processing facility and characterization of isolates by pulsed-field gel electrophoresis and antibiotic resistance profiles. J. Food Prot. 65, 266–273.

Cody, A.J., McCarthy, N.D., Jansen van Rensburg, M., Isinkaye, T., Bentley, S.D., Parkhill, J., Dingle, K.E., Bowler, I.C.J.W., Jolley, K.A., Maiden, M.C.J., 2013. Real-time genomic epidemiological evaluation of human *Campylobacter* isolates by use of whole-genome multilocus sequence typing. J. Clin. Microbiol. 51, 2526–2534.

Colles, F.M., Maiden, M.C.J., 2012. *Campylobacter* sequence typing databases: applications and future prospects. Microbiology 158, 2695–2709.

Corry, J.E.L., Post, D.E., Colin, P., Laisney, M.J., 1995. Culture media for the isolation of campylobacters. Int. J. Food Microbiol. 26, 43–76.

Crim, S.M., Griffin, P.M., Tauxe, R., Marder, E.P., Gilliss, D., Cronquist, A.B., Cartter, M., Tobin-D'Angelo, M., Blythe, D., Smith, K., Lathrop, S., Zansky, S., Cieslak, P.R., Dunn, J., Holt, K.G., Wolpert, B., Henao, O.L., 2015. Preliminary incidence and trends of infection with pathogens transmitted commonly through food—foodborne diseases active surveillance network, 10 US Sites, 2006–2014. Morb. Mortal. Wkly. Rep. 64 (18), 495–499.

Dearlove, B.L., Cody, A.J., Pascoe, B., Méric, G., Wilson, D.J., Sheppard, S.K., 2016. Rapid host switching in generalist *Campylobacter* strains erodes the signal for tracing human infections. ISME J. 10. 721–729.

Deckert, A.E., Taboada, E., Mutschall, S., Poljak, Z., Reid-Smith, R.J., Tamblyn, S., Morrell, L., Seliske, P., Jamieson, F.B., Irwin, R., Dewey, C.E., Boerlin, P., McEwen, S.A., 2014. Molecular epidemiology of *Campylobacter jejuni* human and chicken isolates from two health units. Foodborne Pathog. Dis. 11, 150–155.

Dingle, K.E., Colles, F.M., Ure, R., Wagenaar, J.A., Duim, B., Bolton, F.E., Fox, A.J., Wareing, D.R., Maiden, M.C., 2002. Molecular characterization of *Campylobacter jejuni* clones: a basis for epidemiologic investigation. Emerg. Infect. Dis. 8 (9), 949–955.

Dingle, K.E., McCarthy, N.D., Cody, A.J., Peto, T.E., Maiden, M.C., 2008. Extended sequence typing of *Campylobacter* spp., United Kingdom. Emerg. Infect. Dis. 14 (10), 1620–1622.

Eucker, T.P., Konkel, M.E., 2012. The cooperative action of bacterial fibronectin-binding proteins and secreted proteins promote maximal *Campylobacter jejuni* invasion of host cells by stimulating membrane ruffling. Cell Microbiol. 14, 226–238.

Fenchel, T., Finlay, B., 2008. Oxygen and the spatial structure of microbial communities. Biol. Rev. Camb. Philos. Soc. 83 (4), 553–569.

Food Safety and Inspection Service, 2009. Detection and enumeration method for *Campylobacter jejuni/coli* from poultry rinses and sponge samples. Food Safety and Inspection Service, US Department of Agriculture, Washington, DC. Available from: http://www.fsis.usda.gov/wps/wcm/connect/6bb005d2-2ba0-43dd-aea2-8bf2889757eb/Baseline_Campylobacter_Method.pdf?MOD=AJPERES

Food Safety and Inspection Service, 2013a. Report on *Campylobacter* Testing of Poultry Products—decision to suspend the qualitative test (30 mL). Food Safety and Inspection Service, US Department of Agriculture, Washington, DC. Available from: http://www.fsis.usda.gov/shared/PDF/Campylobacter_Methods_Comparison_Report.pdf

Food Safety and Inspection Service, 2013b. Isolation, identification, and enumeration of method for the enumeration of *Campylobacter jejuni/coli/lari* from poultry rinse, sponge and raw poultry product samples SDA FSIS. Microbiology Laboratory Guidebook (Chapter 41.03). Available from: http://www.fsis.usda.gov/wps/wcm/connect/0273bc3d-2363-45b3-befb-1190c25f3c8b/MLG-41.pdf?MOD=AJPERES

Food Safety and Inspection Service, 2014. Isolation and identification of *Campylobacter jejuni/coli/lari* from poultry rinse, sponge and raw product samples. Microbiology Laboratory Guidebook (MLG) 41.03. Available from: http://www.fsis.usda.gov/wps/wcm/connect/0273bc3d-2363-45b3-befb-1190c25f3c8b/MLG-41.pdf?MOD=AJPERES

Garénaux, A., Jugiau, F., Rama, F., de Jonge, R., Denis, M., Federighi, M., Ritz, M., 2008. Survival of *Campylobacter jejuni* strains from different origins under oxidative stress conditions: effect of temperature. Curr. Microbiol. 56, 293–297.

Gharst, G.A., Oyarzabal, O.A., Hussain, S.K., 2013. Review of current methodologies to isolate and identify *Campylobacter* spp. from foods. J. Microbiol. Methods 95, 84–92.

Gonzalez, I., Grant, K.A., Richardson, P.T., Park, S.F., Collins, M.D., 1997. Specific identification of the enteropathogens *Campylobacter jejuni* and *Campylobacter coli* by using a PCR test based on the ceuE gene encoding a putative virulence determinant. J. Clin. Microbiol. 35, 759–763.

Goossens, H., Pot, B., Vlaes, L., van den Borre, C., van den Abbeele, R., van Naelten, C., Levy, J., Cogniau, H., Marbehant, P., Verhoef, J., 1990. Characterization and description of "*Campylobacter upsaliensis*" isolated from human feces. J. Clin. Microbiol. 28, 1039–1046.

Greig, J.D., Ravel, A., 2009. Analysis of foodborne outbreak data reported internationally for source attribution. Int. J. Food Microbiol. 130, 77–87.

Guccione, E., Leon-Kempis Mdel, R., Pearson, B.M., Hitchin, E., Mulhollandn, F., van Diemen, P.M., Stevens, M.P., Kelly, D.J., 2008. Amino acid-dependent growth of *Campylobacter jejuni*: key roles for aspartase (AspA) under microaerobic and oxygen-limited conditions and identification of AspB (Cj0762), essential for growth on glutamate. Mol. Microbiol. 69, 77–93.

Habib, I., Sampers, I., Uyttendaele, M., et al., 2008. Performance characteristics and estimation of measurement uncertainty of three plating procedures for *Campylobacter* enumeration in chicken meat. Food Microbiol. 25, 65–74.

Habib, I., Uyttendaele, M., De Zutter, L., 2011. Evaluation of ISO 10272:2006 standard versus alternative enrichment and plating combinations for enumeration and detection of *Campylobacter* in chicken meat. Food Microbiol. 28, 1117–1123.

Hilbert, F., Scherwitzel, M., Paulsen, P., Szostak, M.P., 2010. Survival of *Campylobacter jejuni* under conditions of atmospheric oxygen tension with the support of *Pseudomonas* spp. Appl. Environ. Microbiol. 76, 5911–5917.

Hofreuter, D., 2014. Defining the metabolic requirements for the growth and colonization capacity of *Campylobacter jejuni*. Front. Cell Infect. Microbiol. 4, 1–19.

Hofreuter, D., Novik, V., Galan, J.E., 2008. Metabolic diversity in *Campylobacter jejuni* enhances specific tissue colonization. Cell Host Microbe 4, 425–433.

Hutchinson, D.N., Bolton, F.J., 1984. Improved blood-free selective medium for the isolation of *Campylobacter jejuni* from faecal specimens. J. Clin. Pathol. 37, 956–957.

ISO 10272-1:2006. Microbiology of food and animal feeding stuffs—Horizontal method for detection and enumeration of *Campylobacter* spp.—Part 1 Detection method. Available from: http://www.iso.org/iso/iso_catalogue/catalogue_tc/catalogue_detail.htm?csnumber=37091

Jin, S., Song, Y.C., Emili, A., Sherman, P.M., Chan, V.L., 2003. JlpA of *Campylobacter jejuni* interacts with surface-exposed heat shock protein 90alpha and triggers signalling pathways leading to the activation of NF-kappaB and p38 MAP kinase in epithelial cells. Cell Microbiol. 5, 165–174.

Jolley, K.A., Bliss, C.M., Bennett, J.S., Bratcher, H.B., Brehony, C., Colles, F.M., Wimalarathna, H., Harrison, O.B., Sheppard, S.K., Cody, A.J., Maiden, M.C.J., 2012. Ribosomal multilocus sequence typing: universal characterization of bacteria from domain to strain. Microbiology 158, 1005–1015.

Kaakoush, N.O., Castaño-Rodríguez, N., Mitchell, H.M., Man, S.M., 2015. Global epidemiology of *Campylobacter* infection. Clin. Microbiol. Rev. 28 (3), 687–720.

Kelle, K., Pagés, J.M., Bolla, J.M., 1998. A putative adhesin gene cloned from *Campylobacter jejuni*. Res. Microbiol. 149, 723–733.

Kivistö, R.I., Kovanen, S., Haan, A.S., Schott, T., Rahkio, M., Rossi, M., Hänninen, M.-L., 2014. Evolution and comparative genomics of *Campylobacter jejuni* ST-677 clonal complex. Gen. Biol. Evol. 6, 2424–2438.

Konkel, M.E., Garvis, S.G., Tipton, S.L., Anderson, Jr., D.E., Cieplak, Jr., W., 1997. Identification and molecular cloning of a gene encoding a fibronectin-binding protein (CadF) from *Campylobacter jejuni*. Mol. Microbiol. 24, 953–963.

Konkel, M.E., Gray, S.A., Kim, B.J., Garvis, S.G., Yoon, J., 1999. Identification of the enteropathogens *Campylobacter jejuni* and *Campylobacter coli* based on the *cad*F virulence gene and its product. J. Clin. Microbiol. 37, 510–517.

Konkel, M.E., Larson, C.L., Flanagan, R.C., 2010. *Campylobacter jejuni* FlpA binds fibronectin and is required for maximal host cell adherence. J. Bacteriol. 192, 68–76.

Kovanen, S.M., Kivistö, R.I., Rossi, M., Schott, T., Kärkkäinen, U.-M., Tuuminen, T., Uksila, J., Rautelin, H., Hänninen, M.-L., 2014. Multilocus Sequence Typing (MLST) and Whole-Genome MLST of *Campylobacter jejuni* isolates from human infections in three districts during a seasonal peak in Finland. J. Clin. Microbiol. 52, 4147–4154.

Lastovica, A.J., On, S.L.W., Zhang, L., 2014. The Family Campylobacteraceae. In: Rosenberg, E. et al., (Ed.), The prokaryotes—deltaproteobacteria and epsilonproteobacteria. Springer-Verlag, Berlin Heidelberg, pp. 307–335, (Chapter 23).

Lee, C.Y., Tai, C.L., Lin, S.C., Chen, Y.T., 1994. Occurrence of plasmids and tetracycline resistance among *Campylobacter jejuni* and *Campylobacter coli* isolated from whole market chickens and clinical samples. Int. J. Food Microbiol. 24, 161–170.

Lévesque, S., Fournier, E., Carrier, N., Frost, E., Arbeit, R.D., Michaud, S., 2013. Campylobacteriosis in urban versus rural areas: a case–case study integrated with molecular typing to validate risk factors and to attribute sources of infection. PLoS One 8, e83731.

Line, J.E., 2001. Development of a selective differential agar for isolation and enumeration of *Campylobacter* spp. J. Food Prot. 64, 1711–1715.

Line, J.E., Bailey, J.S., Berrang, M.E., 2008. Addition of sulfamethoxazole to selective media aids in the recovery of *Campylobacter* spp. from broiler rinses. J. Rapid Methods Autom. Microbiol. 16, 2–12.

Linton, D., Lawson, A.J., Owen, R.J., Stanley, J., 1997. PCR detection, identification to species level, and fingerprinting of *Campylobacter jejuni* and *Campylobacter coli* direct from diarrheic samples. J. Clin. Microbiol. 35, 2568–2572.

Lior, H., Woodward, D.L., Edgar, J.A., LaRoche, L.J., 1981. Serotyping by slide agglutination of *Campylobacter jejuni* and epidemiology. Lancet 2 (8255), 1103–1104.

Liu, L., Hussain, S.K., Miller, R.S., Oyarzabal, O.A., 2009. Research note: efficacy of mini VIDAS for the detection of *Campylobacter* spp from retail broiler meat enriched in Bolton broth with or without the supplementation of blood. J. Food Prot. 72, 2428–2432.

Luber, P., 2009. Cross-contamination versus undercooking of poultry meat or eggs—which risks need to be managed first? Int. J. Food Microbiol. 134, 21–28.

Luber, P., Brynestad, S., Topsch, D., Scherer, K., Bartelt, E., 2006. Quantification of campylobacter species cross-contamination during handling of contaminated fresh chicken parts in kitchens. Appl. Environ. Microbiol. 72, 66–70.

Maiden, M.C., Bygraves, J.A., Feil, E., Morelli, G., Russell, J.E., Urwin, R., Zhang, Q., Zhou, J., Zurth, K., Caugant, D.A., Feavers, I.M., Achtman, M., Spratt, B.G., 1998. Multilocus sequence typing: a portable approach to the identification of clones within populations of pathogenic microorganisms. Proc. Natl. Acad. Sci. USA 95 (6), 3140–3145.

Maiden, M.C.J., van Rensburg, M.J.J., Bray, J.E., Earle, S.G., Ford, S.A., Jolley, K.A., McCarthy, N.D., 2013. MLST revisited: the gene-by-gene approach to bacterial genomics. Nat. Rev. Microbiol. 11, 728–736.

Miller, R.S., Miller, W.G., Behringer, M., Hariharan, H., Matthew, V., Oyarzabal, O.A., 2010. DNA identification and characterization of *Campylobacter jejuni* and *Campylobacter coli* isolated from cecal samples of chickens in Grenada. J. Appl. Microbiol. 108, 1041–1049.

Millman, C., Rigby, D., Edward-Jones, G., Lighton, L., Jones, D., 2014. Perceptions, behaviours and kitchen hygiene of people who have and have not suffered campylobacteriosis: a case control study. Food Cont. 41, 82–90.

Mohran, Z.S., Guerry, P., Lior, H., Murphy, J.R., El-Gendy, A.M., Mikhail, M.M., Oyofo, B.A., 1996. Restriction fragment length polymorphism of flagellin genes of *Campylobacter jejuni* and/or *C. coli* isolated from Egypt. J. Clin. Microbiol. 34, 1216–1219.

Moser, I., Schroeder, W., Salnikow, J., 1997. *Campylobacter jejuni* major outer membrane protein and a 59-kDa protein are involved in binding to fibronectin and INT-407 cell membranes. FEMS Microbiol. Lett. 157, 233–238.

Nayak, R., Stewart, T.M., Nawaz MS, 2005. PCR identification of *Campylobacter coli* and *Campylobacter jejuni* by partial sequencing of virulence genes. Mol. Cell Probes 19 (3), 187–193.

Ó Cróinín, T., Backert, S., 2012. Host epithelial cell invasion by *Campylobacter jejuni*: trigger or zipper mechanism? Front. Cell Infect. Microbiol. 2, 25.

Occhialini, A., Stonnet, V., Hua, J., Camou, C., Guesdon, J.L., Megraud, F., 1996. Identification of strains of *Campylobacter jejuni* and *Campylobacter coli* by PCR and correlation with phenotypic characteristics. In: Newell, D., Kettly, J., Feldman, R. (Eds.), Campylobacters, helicobacters, and related organisms. Plenum Press, New York, NY, pp. 217–219.

Oh, E., McMullen, L., Jeon, B., 2015. High prevalence of hyper-aerotolerant *Campylobacter jejuni* in retail poultry with potential implication in human infection. Front. Microbiol. 6, 1263.

On, S.L.W., 2001. Taxonomy of *Campylobacter*, *Arcobacter*, *Helicobacter* and related bacteria: current status, future prospects and immediate concerns. J. Appl. Microbiol. 90, 1S–15S.

Oyarzabal, O.A., Liu, L., 2010. Significance of sample weight and enrichment ratio on the isolation of *Campylobacter* spp. from retail broiler meat. J. Food Prot. 73, 1339–1343.

Oyarzabal, O.A., Macklin, K.S., Barbaree, J.M., Miller, R.S., 2005. Evaluation of agar plates for direct enumeration of *Campylobacter* spp. from poultry carcass rinses. Appl. Environ. Microbiol. 71, 3351–3354.

Oyarzabal, O.A., Backert, S., Nagaraj, M., Miller, R.S., Hussain, S.K., Oyarzabal, E.A., 2007. Efficacy of supplemented buffered peptone water for the isolation of *Campylobacter jejuni* and *C coli* from broiler retail products. J. Microbiol. Methods 69, 129–136.

Oyarzabal, O.A., Backert, S., Williams, L.L., Lastovica, A.J., Miller, R.S., Pierce, S.J., Vieira, S.L., Rebollo-Carrato, F., 2008. Molecular typing, serotyping and cytotoxicity testing of *Campylobacter jejuni* strains isolated from commercial broilers in Puerto Rico. J. Appl. Microbiol. 105, 800–812.

Oyarzabal, O.A., Williams, A., Zhou, P., Samadpour, M., 2013. Improved protocol for isolation of *Campylobacter* spp from retail broiler meat and use of pulsed field gel electrophoresis for the typing of isolates. J. Microbiol. Methods 95, 76–83.

Oyarzabal, O.A., Backert, S., 2016. Varying pathogenicity of *Campylobacter jejuni* isolates. In: Gurtler, J.B., Doyle, M.P., Kornacki, J.L. (Eds.), Foodborne Pathogens: Virulence Factors and Host Susceptibility. Springer Publishing, New York, NY.

OzFoodNet Working Group, 2012. Monitoring the incidence and causes of diseases potentially transmitted by food in Australia: annual report of the OzFoodNet network, 2010. Commun. Dis. Intell. Q. Rep. 36, E213–E241.

Pan, M.J., Van Staden, J., 1998. The use of charcoal in in vitro culture. A review. Plant Growth Regul. 26, 155–163.

Penner, J.L., Hennessy, J.N., 1980. Passive hemagglutination technique for serotyping *Campylobacter fetus* subsp *jejuni* on the basis of soluble heat-stable antigens. J. Clin. Microbiol. 12, 732–737.

Persson, S., Olsen, K.E.P., 2005. Multiplex PCR for identification of *Campylobacter coli* and *Campylobacter jejuni* from pure cultures and directly on stool samples. J. Med. Microbiol. 54, 1043–1047.

Platts-Mills, J.A., Liu, J., Gratz, J., Mduma, E., Amour, C., Swai, N., Taniuchi, M., Begum, S., Penataro Yori, P., Tilley, D.H., Lee, G., Shen, Z., Whary, M.T., Fox, J.G., McGrath,

M., Kosek, M., Haque, R., Houpt, E.R., 2014. Detection of *Campylobacter* in stool and determination of significance by culture, enzyme immunoassay, and PCR in developing countries. J. Clin. Microbiol. 52, 1074–1080.

Potturi-Venkata, L.-P., Backert, S., Vieira, S.L., Oyarzabal, O.A., 2007. Evaluation of logistic processing to reduce cross-contamination of commercial broiler carcasses with *Campylobacter* spp. J Food Protect. 2549–2554.

Ray, M., Schwartz, D.C., 2014. Pulsed-field gel electrophoresis and the molecular epidemiology of foodborne pathogens. In: Oyarzabal, O.A., Kathariou, S. (Eds.), DNA methods in food safety: molecular typing of foodborne and waterborne bacterial pathogens. Wiley Blackwell, Chichester, UK.

Reuter, M., Mallett, A., Pearson, B.M., van Vliet, A.H., 2010. Biofilm formation by *Campylobacter jejuni* is increased under aerobic conditions. Appl. Environ. Microbiol. 76, 2122–2128.

Revez, J., Llarena, A.-K., Schott, T., Kuusi, M., Hakkinen, M., Kivistö, R., Hänninen, M.-L., Rossi, M., 2014. Genome analysis of *Campylobacter jejuni* strains isolated from a waterborne outbreak. BMC Genomics, 15, 768.

Ribot, E.M., Fitzgerald, C., Kubota, K., Swaminathan, B., Barrett, T.J., 2001. Rapid pulsed field gel electrophoresis protocol for subtyping of *Campylobacter jejuni*. J. Clin. Microbiol. 39, 1889–1894.

Rodrigues, R.C., Pocheron, A.-L., Hernould, M., Haddad, N., Tresse, O., Cappelier, J.-M., 2015. Description of *Campylobacter jejuni* Bf, an atypical aero-tolerant strain. Gut Pathog. 7, 30.

Sears, A., Baker, M.G., Wilson, N., Marshall, J., Muellner, P., et al., 2011. Marked campylobacteriosis decline after interventions aimed at poultry. N. Z. Emerg. Infect. Dis. 17, 1007–1015.

Seliwiorstow, T., Baré, J., Verhaegen, B., Uyttendaele, M., de Zutter, L., 2014. Evaluation of a new chromogenic medium for direct enumeration of *Campylobacter* in poultry meat samples. J. Food Prot. 77, 2111–2114.

Sellars, M.J., Hall, S.J., Kelly, D.J., 2002. Growth of *Campylobacter jejuni* supported by respiration of fumarate, nitrate, nitrite, trimethylamine-*N*-oxide, or dimethyl sulfoxide requires oxygen. J. Bacteriol. 184, 4187–4196.

Smith, S., Meade, J., McGill, K., Gibbons, J., Bolton, D., Whyte, P., 2015. Restoring the selectivity of modified charcoal cefoperazone deoxycholate agar for the isolation of *Campylobacter* species using tazobactam, a β-lactamase inhibitor. Int. J. Food. Microbiol. 210, 131–135.

Snelling, W.J., McKenna, J.P., Lecky, D.M., Dooley, J.S., 2005. Survival of *Campylobacter jejuni* in waterborne protozoa. Appl. Environ. Microbiol. 71, 5560–5571.

Somerville, G.A., Proctor, R.A., 2013. Cultivation conditions and the diffusion of oxygen into culture media: the rationale for the flask-to-medium ratio in microbiology. BMC Microbiol. 13, 9.

Speegle, L., Miller, M.E., Backert, S., et al., 2009. Use of cellulose filters to isolate *Campylobacter* spp. from naturally contaminated retail broiler meat. J. Food Prot. 72, 2592–2596.

Sproston, E.L., Carrillo, C., Boulter-Bitzer, J., 2014. The quantitative and qualitative recovery of *Campylobacter* from raw poultry using USDA and Health Canada methods. Food Microbiol. 44, 258–263.

Swaminathan, B., Barrett, T.J., Hunter, S.B., Tauxe, R.V., CDC PulseNet Task Force, 2001. PulseNet: the molecular subtyping network for foodborne bacterial disease surveillance, United States. Emerg. Infect. Dis. 7, 382–389.

Taboada, E.N., Acedillo, R.R., Carrillo, C.D., Findlay, W.A., Medeiros, D.T., Mykytczuk, O.L., Roberts, M.J., Valencia, C.A., Farber, J.M., Nash, J.H.E., 2004. Large-scale comparative genomics meta-analysis of *Campylobacter jejuni* isolates reveals low level of genome plasticity. J. Clin. Microbiol. 42, 4566–4576.

Taboada, E.N., Ross, S.L., Mutschall, S.K., Mackinnon, J.M., Roberts, M.J., Buchanan, C.J., Kruczkiewicz, P., Jokinen, C.C., Thomas, J.E., Nash, J.H.E., Gannon, V.P.J., Marshall, B., Pollari, F., Clark, C.G., 2012. Development and validation of a comparative genomic fingerprinting method for high-resolution genotyping of *Campylobacter jejuni*. J. Clin. Microbiol. 50, 788–797.

Tenover, F.C., Williams, S., Gordon, K.P., Nolan, C., Plorde, J.I., 1985. Survey of plasmids and resistance factors in *Campylobacter jejuni* and *Campylobacter coli*. Antimicrob. Agents Chemother. 27, 37–41.

Teramura, H., Iwasaki, M., Ogihara, H., 2015. Development of a novel chromogenic medium for improved *Campylobacter* detection from poultry samples. J. Food Prot. 78, 1624–1769.

Vandamme, P., Dewhirst, F.E., Paster, B.J., On, S.L.W., 2005. Family I: Campylobacteraceae. Brenner, D.J., Krieg, N.R., Staley, J.T. (Eds.), Bergey's manual of systematic bacteriology, vol. 2, second ed. Springer, New York, NY, pp. 1145–1160, Part C.

Wagley, S., Newcombe, J., Laing, E., Yusuf, E., Sambles, C.M., Studholme, D.J., La Ragione, R.M., Titball, R.W., Champion, O.L., 2014. Differences in carbon source utilization distinguish *Campylobacter jejuni* from *Campylobacter coli*. BMC Microbiol. 14, 262.

Winstantley, C., Haldenby, S., Bronowski, C., Nelson, C., Hertz-Fowler, C., Kenny, J., Chaudhuri, R., O'Brien, S., Williams, N., Hall, N., 2015. Application of Whole Genome Sequencing to Fully Characterise Campylobacter Isolates From Infectious Intestinal Disease 1 and Infectious Intestinal Disease 2 Studies (No. FS101072). UK Food Standards Agency, London.

Zhou, P., Oyarzabal, O.A., 2015. Application of Pulsed Field Gel Electrophoresis to Type *Campylobacter jejuni*. In: Jordan, K., Dalmasso, M. (Eds.), Methods in molecular biology, pulsed field gel electrophoresis. Humana Press, a part of Springer Science + Business Media, Totowa, pp. 139–156, 1301.

Zhou, P., Hussain, S.K., Liles, M.R., Arias, C.R., Backert, S., Kieninger, J.R., Oyarzabal, O.A., 2011. A simplified and cost-effective enrichment protocol for the isolation of *Campylobacter* spp from retail broiler meat without microaerobic incubation. BMC Microbiol. 11, 175.

Campylobacter epidemiology—sources and routes of transmission for human infection

Diane G. Newell*,**, Lapo Mughini-Gras*,†,
Ruwani S. Kalupahana§, Jaap A. Wagenaar*,‡,¶

*Department of Infectious Diseases and Immunity, Faculty of Veterinary Medicine, Utrecht University, Utrecht, The Netherlands; **School of Veterinary Medicine, Faculty of Health and Medical Sciences, University of Surrey, Guildford, United Kingdom; †National Institute for Public Health and the Environment (RIVM), Centre for Infectious Disease Control (CIb), Bilthoven, The Netherlands; ‡Central Veterinary Institute of Wageningen UR, Lelystad, The Netherlands; §Department of Veterinary Public Health and Pharmacology, Faculty of Veterinary Medicine and Animal Science, University of Peradeniya, Peradeniya, Sri Lanka; ¶WHO-Collaborating Center for Campylobacter and OIE-Reference Laboratory for Campylobacteriosis, Utrecht/Lelystad, The Netherlands

5.1 INTRODUCTION

Following the routine introduction of simple culture techniques (Skirrow, 1977), it was rapidly established by the early 1980s that the thermophilic *Campylobacter* spp. (*Campylobacter jejuni* and its close relative, *Campylobacter coli*), were a significant cause of diarrhea in humans. Few countries at that time had surveillance of intestinal infectious disease (IID). However, IID had been monitored in England and Wales for many years, and accumulated data quickly determined that, even given the uptake of the new laboratory techniques, campylobacteriosis was increasing year on year, becoming a significant public health burden, with substantial costs. As a similar picture appeared in other industrialized countries, major research efforts were directed toward identifying the sources of infection, and the routes of transmission of human campylobacteriosis. Early epidemiological studies looking for reservoirs established that campylobacters were ubiquitous in the environment, and recoverable from most livestock and pets, as well as foods of animal origin. Nevertheless, these early studies suggested that the food production chain was the most likely source of human infections.

In this chapter, we will review the essential aspects of *Campylobacter* epidemiology, with special emphasis on the approaches used for source attribution of human campylobacteriosis, and on the contribution of potential reservoirs in the food production chain, and elsewhere, to the burden of the disease.

5.2 SOURCE ATTRIBUTION

The attribution of sources and routes of transmission of zoonotic pathogens is an important public health tool, incorporating a growing number of modern methodological approaches and data types. Quantitative estimates of the relative contributions of different sources to the human disease burden can inform risk managers on the priorities for intervention, and enable the implementation and measurement of the impact of interventions, in order to reduce human exposure. A detailed overview of definitions, terminology, and methodologies for source attribution has been presented elsewhere (Pires et al., 2009). Briefly, animals are usually defined as reservoirs or amplifying hosts; the environment and water as potential sources; food and direct contact with animals as examples of transmission routes; meat, dairy, drinking water, etc. as examples of exposures; and swimming in rivers, consumption, and handling of chicken meat, etc. as examples of risk factors. In practice, however, the term "source" is used to refer to any point across the transmission chain.

In the past decade, various approaches have been used to investigate the sources of human campylobacteriosis. Primarily, these approaches comprise epidemiological methods (e.g., analysis of outbreak investigations, and case-control/cohort studies), microbiological methods (e.g., microbial subtyping, and comparative assessment of exposure), or intervention studies, including "natural experiments." Early epidemiological studies for *Campylobacter* were hampered by the ubiquitous distribution of the organism, and the lack of harmonized and stable strain tracing methods. Even today, when genomic analysis is almost routine, the plasticity of the *Campylobacter* genome prevents the timely tracing of specific strains through zoonotic transmission routes. Thus, our current knowledge is founded on the accumulated data from epidemiological, microbiological and intervention studies.

5.2.1 EPIDEMIOLOGICAL METHODS FOR SOURCE ATTRIBUTION

Given the sporadic nature of *Campylobacter* infections in humans (Kapperud et al., 2003), source attribution based on outbreak investigations has had limited value. This is largely because, unlike for salmonellosis (Wagenaar et al., 2013), campylobacteriosis outbreaks are rarely reported. However, they may be more frequent than initially thought (EFSA, 2010c). Outbreak data in Europe during 2005–06 have been analyzed to infer the likely sources (Pires et al., 2010). Although most outbreaks (~64%) were not attributable to known sources, ~12% were attributed to meat products in general, and ~10% specifically to chicken meat. At the individual case level within these outbreaks, about one-third (36%) remained unattributed, but most (~44%) cases were attributed to foreign travel, 17% to contaminated water, and 10% to chicken meat. Therefore, on the basis of outbreak data, chicken meat was the major foodborne source.

Case-control studies have been the most successful epidemiological approach to identify the sources attributable to sporadic cases of campylobacteriosis. Such studies have been conducted in various countries, and a metaanalysis of the data indicates that the handling and consumption of chicken meat is a major risk factor for human campylobacteriosis (Domingues et al., 2012). Other commonly identified risk factors include the consumption of unpasteurized dairy products, eating at restaurants, direct contact with dogs, especially puppies, and with livestock, as well as foreign travel (Studahl and Andersson, 2000; Mughini Gras et al., 2012, 2013; Stafford et al., 2007; Doorduyn et al., 2010; Friedman et al., 2004; Gallay et al., 2008; Danis et al., 2009; Neal and Slack, 1997). From such extensive case-control studies it is generally considered that 20–30% of campylobacteriosis cases are attributable to the handling, preparation, and consumption of chicken meat (EFSA, 2010c).

Case-control studies can be hampered by a number of factors, including acquired immunity. It is well recognized that repeated exposures (at low doses) to pathogens, such as *Campylobacter*, can lead to the development of specific immunity, sufficient to provide protection against clinically overt disease (Swift and Hunter, 2004), but not necessarily against colonization (Havelaar et al., 2009). This effect may explain why campylobacteriosis is rare in adults in the developing world (given later), and that the regular consumption of chicken (at home) is sometimes identified as a protective, rather than a risk, factor (Friedman et al., 2004).

Treatment with gastric antacids, such as proton-pump inhibitors (PPIs), may also affect case-control studies (Mughini Gras et al., 2012; Doorduyn et al., 2010; Tam et al., 2009). PPIs reduce gastric acidity (increased pH) that could favor the survival of *Campylobacter* during passage through the stomach. Alternatively, a damaged gut lining could predispose individuals toward disease (Neal and Slack, 1997). In the Netherlands, the increase in campylobacteriosis observed during 2003–11 correlated with the number of prescriptions for PPIs in the population (Bouwknegt et al., 2014).

5.2.2 MICROBIAL SUBTYPING METHODS FOR SOURCE ATTRIBUTION

Currently, microbial subtyping is the most common approach for source attribution of human campylobacteriosis. *Campylobacter* strains are phenotypically and genotypically highly diverse, and this diversity can be exploited to develop subtyping strategies, with the aim of tracing the organism back to its reservoir. Unfortunately, this approach has been constrained by the high plasticity of the *Campylobacter* genome, and the frequency of horizontal transfer of genetic material between strains, a fact that generates instability in both phenotype and genotype. For many years, these constraints have prevented direct source attribution using serotyping, phage typing, and the simple molecular typing techniques, previously applied to other enteric pathogens.

In 2001, a standardized multilocus sequence typing (MLST) scheme was developed for *C. jejuni* (Dingle et al., 2001). The MLST approach was initially developed to investigate the evolutionary structure of biological populations, but was subsequently applied to source attribution of some zoonotic pathogens. For *Campylobacter*,

MLST depends on the genetic sequence of, usually, seven house-keeping genes. Because of their functions, such genes are highly conserved, but their sequence can vary with time and evolutionary selection pressures. The application of MLST to large *Campylobacter* strain collections has indicated that some sequence types (STs) are associated with specific reservoirs, in particular with poultry and cattle, and even certain environments. By exploiting these associations through the use of stochastic models, it is possible to infer, with some statistical certainty, the origin of those *Campylobacter* strains derived from human cases (Mughini Gras et al., 2012; McCarthy et al., 2007; Wilson et al., 2008; Mullner et al., 2009b; Sheppard et al., 2009; Strachan et al., 2009).

In 2010, MLST-based source attribution models estimated that 50–80% of *Campylobacter* strains infecting humans originate from chickens, 20–30% from cattle, and the rest from other reservoirs (sheep, pigs, and wild animals) (EFSA, 2010c). Similar conclusions have since been drawn from studies worldwide, and the power of using of such approaches for risk management has been exemplified in New Zealand (Muellner et al., 2011). It is interesting to note that source attribution studies in the Netherlands and Luxembourg indicate that both *C. jejuni* and *C. coli* strains isolated from humans are mainly attributable to chicken. However, the proportion of human *C. coli* infections attributed to pigs is far higher than that of *C. jejuni*, whereas the opposite is true for strains attributed to cattle (Mughini Gras et al., 2012; Mossong et al., 2016).

Although MLST has become the most important tool for generating microbial data for source attribution, like other subtyping techniques, it has shortcomings—including the high costs of sampling, isolation, and typing. In addition, the degree of certainty of source attribution from such data is questionable. Many of the strain collections investigated have skewed populations, largely as a result of limited sampling, and poor recovery. Moreover, a significant proportion of *Campylobacter* strains fall into several very large STs (such as ST-21, ST-45, and ST-828) that have a weak host association, thereby reducing the strength of the case for source attribution (Dearlove et al., 2016). As a result, the use of whole genome sequences is increasingly recommended, with even greater cost and, therefore, strain selectivity. The enormous amount of data generated from such approaches makes it increasingly difficult for nonspecialist epidemiologists to analyze and interpret the information in a meaningful way, and apply this information for control and prevention.

5.2.3 COMBINING BOTH EPIDEMIOLOGICAL AND MICROBIAL METHODS

It is known that case-control studies alone do not suffice to attribute human cases to reservoirs (i.e., amplifying hosts) because they can only trace back the sources of human infections up to the point of exposure (e.g., food items consumed, contact with animals, etc.) that may not correspond to the original reservoirs because of, for instance, crosscontamination, or alternative transmission routes (EFSA, 2010c). To overcome this limitation, the MLST of strains isolated during targeted epidemiological studies has been undertaken (Mughini Gras et al., 2012; Mullner et al., 2009a; Mossong et al., 2016). This combined approach suggests that transmission routes other than foodborne are also important. For example, in a Dutch investigation using strains isolated from a case-control study, infections caused by chicken-associated strains are not only associated with the consumption and handling of chicken meat, but also with contact with individuals suffering from gastrointestinal symptoms (Mughini Gras et al., 2012). This suggests that person-to-person spread of these strains is more frequent than previously thought. Also, ruminant-associated strains are linked not only to the consumption of tripe and barbecued meat, but also to occupational exposure and rural living, suggesting that direct/indirect environmental contact with these animals is important. The consumption of game meat and the use of swimming pools were also significant risk factors for infection with *Campylobacter* strains originating from the environment, especially during springtime. It seems likely that, as more such combined studies are undertaken, the nonmeat-foodborne aspects of campylobacteriosis will become clearer.

5.2.4 NATURAL EXPERIMENTS AND INTERVENTIONS FOR SOURCE ATTRIBUTION

There have been several "epidemiological events" involving poultry that have served as natural experiments of the effect of a major and sudden reduction of the exposure to *Campylobacter* in the human population. In 1999, the crisis following the finding of dioxin in animal feed in Belgium led to the national withdrawal of various poultry products intended for human consumption. Subsequently, there was a drastic reduction in the nationwide consumption of chicken meat, a fact that was associated with a concurrent drop of 40% in campylobacteriosis across Belgium (Vellinga and Van Loock, 2002). Similarly, in 2003, an epidemic of avian influenza (H7N7) hit the Netherlands. To control this epidemic, massive bird culling measures were implemented (~30 million birds culled), and a number of poultry slaughterhouses were closed (Stegeman et al., 2004). The epidemic was associated with simultaneous declines in campylobacteriosis locally and nationally (50 and 30%, respectively) (Friesema et al., 2012). Although sales of poultry meat declined throughout the Netherlands by ~9% during this period, this alone was considered insufficient to account for the reduction in campylobacteriosis. Moreover, this reduction continued far beyond the recovery in poultry meat sales. Overall, the analysis of this natural experiment has suggested that a significant proportion of the observed public health benefit resulted from the reduction in the environmental burden of campylobacters originating from poultry.

Unfortunately, such natural experiments only allow for the retrospective observation of effects. The implementation of national intervention programs to reduce *Campylobacter* on poultry meat, however, allow prospective opportunities to study the effect of reduced exposure to poultry-associated campylobacters. For instance, following interventions in Iceland and New Zealand, the total number of campylobacteriosis cases decreased by 72% (Stern et al., 2003) and 54% (Sears et al., 2011), respectively. In New Zealand, MLST-based source attribution analyses showed that this reduction was largely due to a fall in the cases of poultry-associated campylobacteriosis (Muellner et al., 2011).

5.2.5 TRANSMISSION ROUTES FOR POULTRY-ASSOCIATED CAMPYLOBACTERS NOT INVOLVING THE HANDLING AND CONSUMPTION OF POULTRY MEAT

By 2010, it was apparent that there was a significant discrepancy between the attribution of the handling and consumption of poultry meat as a source of human campylobacteriosis based on case-control studies (20–40%), and that of the chicken reservoir as a whole (50–80%) based on MLST (EFSA, 2010c). This observation has been more recently supported by the analysis of the avian influenza outbreak in the Netherlands (given earlier), and by combining molecular and epidemiological approaches to source attribution (Mughini Gras et al., 2012). Overall, current opinion accepts that, although poultry are the major livestock reservoir for campylobacteriosis, there are various routes of transmission to humans, and the handling and consumption of poultry meat may not be the cause of the majority of such infections.

5.3 FOODBORNE SOURCES OF CAMPYLOBACTERIOSIS

5.3.1 *CAMPYLOBACTER* IN CHICKEN AND OTHER POULTRY

Although the relative importance of poultry meat as the primary cause of campylobacteriosis may be debatable, the importance of chicken as a major reservoir of infection is indisputable. Reflecting this, most public health activities aimed at the reduction of campylobacteriosis have focused on poultry production.

C. jejuni and *C. coli* appear to have evolved to preferentially colonize the avian gut. Most domestic poultry reared for consumption or egg production, including chicken, turkey, geese, ducks, pigeons, and even ostriches, as well as wild birds, are frequently colonized with these *Campylobacter* species. This colonization appears to be asymptomatic, except perhaps in young ostriches, and thus has no obvious cost implication for farmers. The epidemiology of *C. jejuni/coli* colonization in broilers has been comprehensively reviewed elsewhere (Wagenaar et al., 2008). Therefore, only references that contribute additional information will be indicated in this section.

5.3.1.1 Campylobacter *in live broiler chicken*

The most significant data on *Campylobacter* colonization of poultry has been derived from the conventional broiler industry, because this sector has been easily investigated, and is the largest of the poultry industry. Over the past three decades, many studies throughout the industrialized world have reported on the prevalence of broiler colonization with *Campylobacter*, and a huge variation has been observed. However, comparisons between studies and countries are hampered by differences in sampling, culture methodology, and flock management factors. In 2008, a European Union baseline survey of *Campylobacter* colonization of broiler flocks was undertaken, using standardized sampling procedures and detection methods on 10,132 batches of broilers from 561 slaughterhouses, in 28 European countries (EFSA, 2010a). The average prevalence of *Campylobacter*-positive broiler batches was 71.2%, but this varied between countries from 2% to 100%, with the lowest levels being reported in Scandinavian countries. This European study remains the most comprehensive to date, and confirms that each country, and maybe even geographical region, needs to establish their own baseline prevalence before undertaking any intervention.

There are several interesting common features in the outcomes of broiler epidemiological studies. First, prevalence is directly related to the age of the flock. It is now generally accepted that campylobacters are rarely, if ever, vertically transmitted, so that chicks are hatched *Campylobacter*-free. Positive flocks are rarely detected until 2–3 weeks of age. The reason for this is unclear, but maternal immunity (Cawthraw and Newell, 2010) and/or lack of exposure are likely explanations. All birds in a flock rapidly become colonized following the first detection. Experimental challenges with fresh chicken strains show that the dose for successful chicken colonization can be very low (less than 10 cfu), and colonization levels in the cecum increase to over 10^9 cfu/g of cecal contents within 3 days. Coprophagy and feed/water contamination ensure rapid spread of the organism, so that the with-flock prevalence is about 100% within days of the first detection. Effective bird-to-bird transmission is also enhanced by bacterial adaptation to efficient colonization of the chicken gut.

Flock prevalence is also seasonal, with a summer peak that is also geographically distributed, with countries of high latitudes, that is, Scandinavia, having a peak later in the season (EFSA, 2010b). Explanations for these effects include weather patterns (i.e., rainfall and sun/UV levels), allowing/preventing *Campylobacter* survival in the environment, and fly seasons enabling *Campylobacter* transmission from the environment. Flock management systems, such as levels of containment and biosecurity, certainly have an effect on prevalence (Newell et al., 2011). Some farming practices can significantly increase the horizontal transmission of campylobacters from the environment into the flock. Factors such as thinning (the structured reduction of bird numbers prior to flock harvesting), multilivestock farming, and the extensive rearing of flocks can all increase the risk of flock positivity. *C. jejuni* is the most prevalent species recovered from broilers, but *C. coli* prevalence may increase in older birds. Most intensively reared flocks are colonized with a limited number of strains. Frequently, only a single strain can be detected, suggesting limited exposure, competitive exclusion and/or differences in strain colonization potential. Finally, there is increasing evidence that the genetic lineage of broilers can affect susceptibility to colonization with *Campylobacter* (Psifidi et al., 2016), though this has yet to be confirmed in the field situation.

5.3.1.2 Campylobacter *in other birds*

As indicated previously, *C. jejuni/coli*, appear to have a preference for the ecological niche provided by the avian gut. It is not surprising, therefore, that all other poultry can be colonized, including turkeys, ducks, geese, guinea fowl, ostrich, and pigeon, as well as game birds and caged pet birds. The rearing and management of such birds in intensive or enclosed conditions may contribute to bird-to-bird transmission. The issue of wild birds will be addressed later.

5.3.1.3 Campylobacter *in poultry meat*

During processing, campylobacters that have been colonizing the chicken gut can contaminate the carcass surface to a variable extent, but frequently at levels of over 1×10^6 cfu per carcass (EFSA, 2010a). Quantitative risk assessment models indicate that higher levels of contamination generate the greatest risk to consumers (EFSA, 2010c). This contamination arises from fecal leakage, gut tissue damage, or feathers soiled during rearing or transport to the abattoir (reviewed by Jacobs-Reitsma and Lyhs, 2008). Several stages of processing at the abattoir contribute to this contamination, including scolding, defeathering, and evisceration. Although the risk of carcass contamination is higher from colonized birds, processing also contaminates the abattoir equipment, potentially transferring organisms, usually at low numbers, onto the carcasses of uncolonized birds subsequently entering the processing line. Surveys of contaminated broiler carcasses have been undertaken at the abattoir, after processing, and at retail. Comparison of the data from such surveys is confounded by unstandardized approaches to sampling and culture. Analysis of the 2008 baseline survey undertaken in Europe (EFSA, 2010b) clearly showed that levels of contamination varied between countries, and between slaughterhouses.

Because of their fastidious nature, high numbers of campylobacters contaminating poultry carcasses do not multiply, but can survive through to retail, and constitute a risk to customers from handling and crosscontamination in the kitchen. *C. jejuni/ coli* are fragile organisms, highly susceptible to the toxic effects of atmospheric oxygen, dehydration, temperature, and many chemicals, so the numbers of organisms surviving on carcasses to retail is highly dependent on the initial level of contamination, and the postprocessing methods employed. Washing in appropriately chlorinated water, or water containing other decontaminating chemicals, such as organic acids, significantly reduce the surface contamination levels, and can be used for public health interventions. However, regulations for the postprocessing chemical treatment of poultry meats vary between countries. Freezing and air-drying can reduce levels on the carcass by factors of several hundredfolds. However, survival is enhanced if the organisms are located within the feather follicles. Unfortunately, the poultry industry approach, which increasingly reduces the time between slaughter and retail, and introduces storage and packaging methods aimed at improving meat shelf life (cool, dark, and moist conditions wrapped in plastic largely excluding the normal atmosphere), enables long term survival of *Campylobacter* on poultry meat, ensuring the viability of the organism at the point of sale. It should be also noted that the surface of poultry meat packaging can itself be contaminated with campylobacters (Harrison et al., 2001), constituting a risk to customers at the retail shelf level.

The greatest proportion of campylobacters are located on the poultry skin or carcass surface, and easily destroyed by cooking. Organisms can be detected in poultry muscle, but the levels are generally low. However, chicken liver can be highly contaminated (Jacobs-Reitsma and Lyhs, 2008), and constitute a particular risk if consumed undercooked.

5.3.1.4 Campylobacter *in laying hens and eggs*

Like other commercial breeds of *Gallus gallus*, table-egg laying hens are frequently colonized with campylobacters. Viable organisms are recoverable from the oviducts of laying hens, and egg shells can be contaminated, presumably from feces, but culturable campylobacters have been rarely recovered from egg contents (Wagenaar et al., 2008), suggesting that eggs are rarely, if ever, a source of campylobacteriosis. This is consistent with outbreak data, the age-related flock colonization data, and the lack of colonization in experimental control groups of chicken.

5.3.2 *CAMPYLOBACTER* IN RUMINANTS

Source attribution studies indicate that domestic ruminants (cattle, sheep, and goats) are the second most important reservoir of *C. jejuni/coli* strains, causing up to 33–38% of human campylobacteriosis cases (Jonas et al., 2015; Mossong et al., 2016; Sheppard et al., 2009).

Campylobacter-colonized ruminants pose a risk through the consumption of contaminated meat, offal, and dairy products and, surprisingly, studies in Luxembourg (Mossong et al., 2016) and the Netherlands (Doorduyn et al., 2010) report that consuming beef (both in and out of the home) is a particular risk factor for *C. coli* infection. However, the main risk of campylobacteriosis from ruminants seems to be through environmental contamination, which will be discussed later.

5.3.2.1 Campylobacter *in live ruminants*

C. jejuni/coli gastrointestinal tract carriage is usually asymptomatic in ruminants. However, both species may cause abortion in sheep and cows. Many prevalence surveys have been undertaken. The absence of standardized approaches to sampling strategy and detection approaches precludes direct comparison of data. However, in Finland and the United Kingdom, structured surveys of fecal samples from cattle at slaughter reported that 31.1 and 54.6% of animals, respectively, carried *Campylobacter* spp. (Hakkinen et al., 2007; Milnes et al., 2008). For small ruminants (sheep and goats), a structured survey undertaken at slaughter in the United Kingdom reported that 43.8% of sheep carried thermophilic campylobacters (Milnes et al., 2008), a fact that was significantly different from carriage in cattle, and is similar to data (32.8%) from sheep and goats at slaughter in Greece (Lazou et al., 2014). However, carriage on farms may be lower for both cattle and sheep (22 and 25%, respectively) than at the abattoir (Rotariu et al., 2009), perhaps reflecting the effect of stress of transport and lairage on the gut microbiome. Several studies have indicated that carriage is generally higher in calves than older animals (Sasaki et al., 2013; Sato et al., 2004; Nielsen, 2002), and may be slightly lower in dairy cows (Johnsen et al., 2006; Merialdi et al., 2015). However, fecal shedding is intermittent (Jones et al., 1999; Stanley et al., 1998a,b), and may be increased by management factors such as holding animals in feedlots (Besser et al., 2005). In cattle and small ruminants, the reports of the ratio of *C. jejuni* to *C. coli* recovered vary widely. These differences are probably a reflection of many factors, but especially the sample handling, the isolation, culture and detection methods used, and possibly the age of the animals. In the structured national survey of Milnes et al. (2008), 81% of cattle and 65% of sheep isolates were *C. jejuni*, and the remainder primarily *C. coli*.

The application of manure with broadcast spreaders, indoor housing, feed composition, herd size, private water supply, presence of horses in the farm, and the accessibility of feed to birds, have all been identified as significant risk factors for *C. jejuni* carriage in dairy cattle (Ellis-Iversen et al., 2009; Wesley et al., 2000), while a high number of female animals on the farm has been identified as a risk factor in beef cattle (Hoar et al., 2001).

There are generally two seasonal peaks in *Campylobacter* shedding in cattle: one in late spring, and the other in late autumn, with some variation among countries (Hakkinen and Hänninen, 2009; Milnes et al., 2008; Sato et al., 2004; Stanley et al., 1998a). These periods roughly coincide with traditional milk flushes and calving periods, as well as with spring transition from winter housing to summer grazing, and the autumn return to winter housing, suggesting that shedding patterns might reflect either hormonal and stress influences to the gut flora, or changes in diet and water sources. However, nowadays most cattle farmers in developed countries calve all year round, so the association of the seasonal peaks with calving requires further clarification (Merialdi et al., 2015). In small ruminants, the pattern of *Campylobacter* shedding is better defined (Jones et al., 1999), at least in the United Kingdom, with a seasonal low in November–December, when sheep are fed on hay and silage, rather than grazing, and a seasonal high coincident with lambing, weaning, and movement onto new pasture.

5.3.2.2 Campylobacter *in foodstuffs of ruminant origin*

As indicated earlier, milk and dairy products are frequently implicated as vehicles of human campylobacteriosis outbreaks. It is generally assumed that this is the result of fecal contamination, but outbreaks have been occasionally traced back to asymptomatic *C. jejuni* mastitis in dairy cattle (Orr et al., 1995). Effective conventional pasteurization kills campylobacters, but raw or incompletely pasteurized milk may contain viable organisms (Fernandes et al., 2015). In developed countries, contamination in bulk-tank milk is usually low, below 1% (Hill et al., 2012; Muehlherr et al., 2003; Ruusunen et al., 2013), but can be as high as 12% in some instances (Bianchini et al., 2014).

Fecal contamination of carcasses can occur during evisceration, removal of the hide, or because of crosscontamination within the abattoir. However, the prevalence of campylobacters on beef at retail is generally low, for example, 4.5% in the United Kingdom (Little et al., 2008), 3.2% in Ireland (Whyte et al., 2004), and 3.5% in New Zealand (Wong et al., 2007). In the same surveys, slightly higher prevalences were found for lamb/mutton at 12.6, 11.8, and 6.9%, respectively. However, *Campylobacter* contamination of ruminant offal, in particular liver, is substantially higher

(Little et al., 2008). The lower *Campylobacter* contamination of red, compared with poultry, meats is assumed to reflect the wetter conditions during poultry processing that would enable bacterial survival on product surfaces.

5.3.3 *CAMPYLOBACTER* IN PIGS AND PORK PRODUCTS

Only about 0.4% of all human campylobacteriosis cases can be attributed to pigs (Mossong et al., 2016), but for *C. coli* infections this contribution increases to 4.4–6% (Sheppard et al., 2009; Mossong et al., 2016). Since pigs are rarely considered an important reservoir for campylobacteriosis, there are few initiatives to control *Campylobacter* in the pork production chain. However, in some countries the high antimicrobial resistance observed in strains from pigs is considered a particular risk to humans (Quintana-Hayashi and Thakur, 2012). Only under exceptional conditions, such as production in areas remote from other pig producing facilities, might *Campylobacter*-free pork production be feasible (Kolstoe et al., 2015).

5.3.3.1 Campylobacter *in the live pig*

Few large surveys on the prevalence of *Campylobacter* in live pigs have been undertaken, but based on cultured rectal samples at the farm or abattoir, about 38.1–63% of pigs carry these organisms (Nathues et al., 2013; Carrique-Mas et al., 2014; Milnes et al., 2008). *Campylobacter* colonization in pigs is asymptomatic. The concentration of *Campylobacter* in porcine fecal samples is up to 1.2×10^7 cfu/g (Abley et al., 2012). In most studies, the vast majority of these bacteria (>90%) are *C. coli*, and the remainder being *C. jejuni* (Milnes et al., 2008; Quintana-Hayashi and Thakur, 2012), so it seems that the porcine gut is a preferred niche for this species. However, one study from Vietnam found *C. jejuni* as the predominant species in pigs (Carrique-Mas et al., 2014). Whether this reflects a geographical, or methodological, difference is unknown. Little is currently known about the risk factors for *Campylobacter* colonization in pigs.

5.3.3.2 Campylobacter *in foodstuffs of porcine origin*

There have been several surveys of *Campylobacter* in pork products, but the approaches and methodologies vary significantly. The European Food Safety Authority (EFSA) annually collects any data submitted from Member States on food contamination. In 2014, it was reported that 6–37% of pig carcasses (from Belgium and Poland, respectively), 2.6–6.8% of meat samples at the processing plant (from Hungary and Portugal, respectively), and 0.27–2% of pork meat at retail (from the Netherlands and Spain, respectively) were *Campylobacter*-positive (EFSA and ECDC, 2015). In surveys from four European countries (United Kingdom, Belgium, Poland, and Italy), the prevalences in pork (meat and chops) at retail was 5.0, 5.0–16.6, 10.6, and 5.7%, respectively (Little et al., 2008; Mattheus et al., 2012; Korsak et al., 2015; Sammarco et al., 2010), but the prevalence was higher in pork offal (Little et al., 2008). Interestingly, the Polish study reported *C. jejuni* as the most prevalent species on pork chops. In the USA, *Campylobacter* prevalence on pork chops was reported to be only 0.3–2% (Noormohamed and Fakhr, 2013; Zhao et al., 2010).

5.3.4 *CAMPYLOBACTER* IN READY-TO-EAT FOODS

Some case-control studies have indicated that ready-to-eat (RTE) foods, such as takeaways, pose a risk for campylobacteriosis (Rodrigues et al., 2000). Most investigations of RTE foods at retail in Europe have failed to find *Campylobacter* spp. (Moore et al., 2002; Meldrum and Ribeiro, 2003). However, a few surveys have identified or recovered campylobacters from cooked meats (Maćkiw et al., 2011; Elson et al., 2004), but at a prevalence of >1%. The likelihood is that crosscontamination of such foods occurs postcooking, rather than survival of the organisms after cooking. Such crosscontamination can easily occur in a kitchen (Luber et al., 2006), and risk managers actively communicate the risks of handling raw and cooked meats together.

Perhaps of more concern is the detection of campylobacters at retail in vegetables produced to be potentially eaten raw. For example, in the Netherlands, 0.83% of endive and 2.7% of oak tree lettuce samples were positive (Wijnands et al., 2014). Campylobacters have also been isolated from leafy greens (Ceuppens et al., 2015), ulam (Khalid et al., 2015), and lettuce (de Carvalho et al., 2013), and have been shown to survive for at least a day and up to 8 days on various fruits and vegetables (Kärenlampi and Hänninen, 2004). An outbreak of campylobacteriosis associated with the consumption of raw peas has been described (Gardner et al., 2011). In some investigations, the eating of raw fruit and vegetables are risk factors for campylobacteriosis (Fullerton et al., 2007; Verhoeff-Bakkenes et al., 2011), although in others (Kapperud et al., 2003; Mughini Gras et al., 2012) their consumption is protective.

The use of irrigation water contaminated by feces from livestock is an obvious candidate source. Another potential transmission route would be the direct contamination of the produce by feces from wild birds. This is supported by the subtyping of isolates from the raw pea outbreak (Gardner et al., 2011). Also, *C. lari* can be one of the contaminating *Campylobacter* species (Losio et al., 2015), and such strains are frequent colonizers of wild birds.

5.4 NONFOODBORNE SOURCES OF CAMPYLOBACTERIOSIS

Source attribution studies have estimated that, relative to poultry, ruminants, and pigs, the general environment may account for up to 5–10% of human campylobacteriosis cases (Mughini Gras et al., 2012; Mossong et al., 2016). Such environmental contribution would comprise all surface water (lakes, rivers, puddles, etc.), soils and air, as well as pets, wildlife, and livestock other than poultry, ruminants, and pigs. The range of environmental sites contaminated with campylobacters has recently been reviewed (Whiley et al., 2013). However, the extent of such contamination has yet to be established, and it should be kept in mind that fecal waste, deposited in the environment, can also be widely disseminated by rain water, wind, animal movement, and flying insects to generate further indirect contact routes. Since *Campylobacter* cannot grow naturally outside the gut of a warm-blooded animal, the general environment can only serve as a vehicle for transmission, and not as an amplification niche. Moreover, the fragility of *Campylobacter* means that survival in the environment will be dependent on exogenous factors, such as temperature, sunlight, and oxygen exposure, absence of moisture, etc. Therefore, it seems reasonable to assume that airborne infection of humans, at least at any distance remote from heavy contamination sources like broiler houses, is unlikely, but that water and soils are likely to be contaminated.

As indicated previously, the level of shedding of *Campylobacter* from colonized livestock depends on a number of factors, including age, stress, changes in diet, and housing conditions, as well as with season, often with a summer peak. Human exposure to potential environmental sources, for example, swimming in a domestic pool, is largely weather-dependent, so not surprisingly the risk of infection with environment-associated strains is also seasonal (Mughini Gras et al., 2012). However, there is also a risk associated with living in a rural environment, especially in young children (Strachan et al., 2009), who might be exposed to a wider range of strains than urban dwellers.

5.4.1 *CAMPYLOBACTER* IN DOGS, CATS, AND HORSES

Up to 87% of dogs are colonized with *Campylobacter* species (see review by Marks et al., 2011). The most prevalent species in dogs is *C. upsaliensis* (Carbonero et al., 2012; Acke et al., 2009). However, *C. jejuni* also colonizes dogs, with a prevalence of 0–45% (Marks et al., 2011). It remains unclear whether *Campylobacter* causes disease in dogs. One case-control study of diarrheic and healthy dogs showed 97 and 58% *Campylobacter* prevalence, respectively, by PCR, while a culture-based study reported no significant difference between such groups (Chaban et al., 2010; Stavisky et al., 2011). The evidence for an increased risk of campylobacteriosis for dog owners is also debatable, with some studies indicating a risk (Mughini Gras et al., 2013), and others not (Rodrigues et al., 2000).

In cats, up to 42.9% of animals are reported to be *Campylobacter* positive, with *C. helveticus* as the predominant colonizing species (Acke et al., 2009; Wieland et al., 2005). Data from horses is scarce, but the reported prevalence is low (<5%) (Moriarty et al., 2015; Roug et al., 2013). The risk of direct contact between *Campylobacter*-colonized companion animals, especially to people with little prior exposure, at community events such as country fairs, petting farms, and even in residential care, has yet to be understood.

5.5 SURFACE WATERS, SOILS

Surface waters are a particular problem, and can be considered "sinks" that collect *Campylobacter* strains from various animal reservoirs. Recreational and alluvial waters are frequently contaminated with campylobacters (Arnone and Walling, 2007), and water-related activities, such as swimming, and even children's paddling pools (Sawabe et al., 2015), can constitute a risk of infection. The recovery of *Campylobacter* from surface waters is often indicative of recent contamination with wastewater effluents or agricultural runoff (Jones, 2001). For example, in Ontario, Canada, in 2000, well water serving the town of Walkerton was contaminated by *Escherichia coli* O157:H7, and *C. jejuni* from cattle waste washed out from local farms by heavy rainfall, resulting in at least seven deaths, and over 2000 illnesses (Clark et al., 2003). The presence of wild birds, including water fowl, is also a source of surface water contamination, with an associated risk of campylobacteriosis (Mullner et al., 2009b).

The survival of *Campylobacter* in surface waters is dependent on a number of factors, including the hours of sunlight, temperatures, and water quality (reviewed by Whiley et al., 2013). There is increasing evidence for the presence of environmentally adapted *Campylobacter* strains, such as ST45, known to be particularly widespread in the environment (French et al., 2005; Sopwith et al., 2008). Such strains appear to show an enhanced fitness outside the host, presumably as a result of the evolution of bacterial stress mechanisms to improve survival (Sopwith et al., 2008). Survival in the environment may also be enhanced by uptake of the bacteria in protozoa (Trigui et al., 2016; Olofsson et al., 2015), or by the development of biofilms (Pascoe et al., 2015).

5.5.1 POTABLE WATER

The main evidence for drinking water as a source of campylobacteriosis is from outbreak, rather than sporadic, cases. As indicated earlier, such outbreaks can be very large. Interestingly, many occur in countries of northern latitudes, such as in Scandinavia, where survival might be enhanced by reduced sunlight and cooler temperatures. In such countries, drinking water is also recognized as a risk factor for broiler colonization (Newell et al., 2011).

Campylobacters are rarely isolated from potable water supplies, unless there is a breakdown in treatment plants (Jones, 2001). However, in rural areas where drinking water may be supplied from wells, or by rainwater collection, the opportunities for contamination are likely to be high, especially when water is stored in tanks open to wild birds and animals.

5.5.2 WILD BIRDS

Wild birds, of many families, are frequently colonized with *Campylobacter* strains (Colles et al., 2011; Griekspoor et al., 2013), some of which demonstrate distinct host specificity. It is estimated that about 3.5% of human campylobacteriosis cases are attributable to wild bird strains (Cody et al., 2015), with a peak in the summer months, but, in children under 5 years of age living in rural areas, this risk could be as high as ~24% of infections (Strachan et al., 2009). In addition to surface water, other transmission routes for wild bird-associated *Campylobacter* strains are the exposure to fecal material in children's playgrounds (French et al., 2009), pecking of milk-bottle tops (Neal and Slack, 1997), and outdoor leisure or work activities (Strachan et al., 2009). However, stochastic models based on MLST suggest that *Campylobacter* strains from wild birds pose a lower risk to humans, relative to those from domestic poultry. The reason for this is unknown, but host specialization may play an important role.

5.5.3 INSECTS

There is increasing evidence that flying insects, such as the house fly (*Musca domestica*), are transmission vectors, able to carry campylobacters from fecal material in the environment to another host (Förster et al., 2009) that could be chickens or, possibly, humans. A recent risk assessment (Evers et al., 2016) suggests that the importance of this exposure route to humans should be further investigated.

5.6 *CAMPYLOBACTER* EPIDEMIOLOGY IN THE DEVELOPING WORLD

Much of our understanding of the epidemiology of campylobacteriosis comes from investigations within the industrialized world. Travelers from industrialized countries often have a higher risk of acquiring campylobacteriosis, and this risk seems to be particularly high for Western travelers going to Asia, Africa, Latin America, and the Caribbean (Mughini-Gras et al., 2014). It is assumed, but not proven, that this risk reflects a greater environmental, as well as foodborne, burden in these geographical regions, as well as the possibility of exposure to "exotic" strains previously unencountered in the home environment.

The awareness, and monitoring, of campylobacteriosis in the developing world is generally poor, and this has been highlighted by the World Health Organization (WHO) (http://www.who.int/foodsafety/publications/campylobacteriosis/en/). One well-recognized feature of human epidemiology in the developing world is that colonization with *Campylobacter* is common but, in adults, largely asymptomatic. In young children, however, colonization is often associated with disease. One explanation for this is that adults in developing countries have been frequently exposed to *Campylobacter* through food and the environment, and consequently have developed immunity that protects from disease but not colonization (Havelaar et al., 2009).

In terms of World Bank classification, developing countries denote all low- and middle-income countries. In such countries, livestock production (especially of poultry and pigs) is largely extensive, frequently on widely distributed small-holder farms, or in backyards (Gilbert et al., 2015), which minimizes labor and cost inputs. The little information to date on potential sources of campylobacteriosis in the developing world comes primarily from poultry. Small and large scale commercial poultry farms mostly use a deep litter open-house system where broilers are constantly in contact with external environment, wild animals, and flies. However, village or family based extensive poultry production is also common, and in these cases biosecurity is rarely feasible or practical (Conan et al., 2012). For example, even 1-day-old chicks are in close contact with adult birds likely to be already colonized with *Campylobacter* (Kalupahana et al., 2013). Extensive farming methods in developed countries are known to increase the risk of *Campylobacter* colonization in broilers (Allen et al., 2011). Therefore, high prevalences of *Campylobacter* colonization of broilers at slaughter in developing countries should be expected. In studies from several developing countries, for example, China (Ma et al., 2014), levels of over

70% have been reported. However, much lower levels have been reported elsewhere, for example, 31.9% in Vietnam (Carrique-Mas et al., 2014). Surprisingly, this range of prevalence levels would seem to be little different from those reported in the developed world, but bird age, sampling strategies, culture methodologies, and microbiological competence could all contribute to this observation.

In developing countries, the production of poultry meat is a combination of mechanized abattoir processing, which is to some extent regulated, and an unregulated informal sector, often using live bird markets (alternatively known as wet markets, or pluck shops). In some countries, for example, India, the proportion of chicken meat sold from freshly culled chickens at live bird markets is over 90% (Parkar et al., 2013) (http://gain.fas.usda.gov/Recent%20GAIN%20Publications/Poultry%20and%20Poultry%20Products%20Annual%202015_New%20Delhi_India_9-30-2015.pdf). Local consumer preference determines the type of retail product sold. In some countries, hand-slaughtered chicken is considered fresher and healthier, but in others, such as Sri Lanka, consumers prefer mechanically processed meat. Evidence is increasing for the lower hygiene standards in the preparation of carcasses produced in wet markets. For example, in Yangzhou, China (Huang et al., 2016), although the prevalence of *Campylobacter*-contaminated carcasses retailed from wet markets and supermarkets was about the same (63.3% vs. 66.7%), the levels of *Campylobacter* contamination on the carcasses was significantly higher from the former.

To date, there has been little further characterization of the campylobacters isolated in the developing world. A few studies have reported a higher proportion of *C. coli* isolates being recovered in developing countries (Ansari-Lari et al., 2011; Salihu et al., 2009), than anticipated from surveys in developed countries. In addition, high levels of antimicrobial resistance have been reported in some *Campylobacter* isolates from developing countries (Ansari-Lari et al., 2011; Parkar et al., 2013; Carrique-Mas et al., 2014).

Overall, the situation regarding *Campylobacter* epidemiology in developing countries remains unclear. Surveillance for infectious intestinal diseases is generally poor in these countries. Nevertheless, recent analysis indicates that *Campylobacter* spp. cause the highest bacteria-attributable burden of diarrhea in the first and second years of life, and that to date this burden has been greatly underestimated (Platts-Mills et al., 2015). The sources of these infections are as yet unidentified, but food, and especially poultry meat, would seem to have an important role.

5.7 CONCLUSIONS

Given the successful source attribution approaches previously applied to salmonellosis, our inability to identify the routes of transmission for campylobacteriosis during the last two decades of the 20th century came as a considerable surprise. The control and prevention of *Salmonella* infection relied heavily on outbreak investigations, case-control studies, recovery of isolates from putative sources, and typing of those isolates to track strains from reservoirs to man. In this chapter, we have reported that

Campylobacter was less amenable to these approaches, for a number of reasons. In particular, outbreaks were relatively infrequent and poorly reflected in case-control studies, which in themselves were hampered by the self-limiting nature and extended incubation period for the disease. Tracing the strains through the transmission route was hampered by the fragility of the organisms in the environment. Moreover, despite the considerable heterogeneity of *C. jejuni/coli* strains, the population structure was very different from that of the salmonellas. This meant that the simple, largely serologically based, typing methods, so extensively and successfully used to trace individual *S. enterica* strains through the environment, were unavailable to *Campylobacter* epidemiology. In an attempt to "force campylobacters to comply with the needs of the epidemiologists," increasingly sophisticated typing methods were developed and applied worldwide to the growing collections of *Campylobacter* strains from humans and their environment. Over the past 15 years, the development and application of MLST has enabled the accumulation of evidence worldwide for overlapping animal and human *Campylobacter* populations, and statistical approaches have allowed disease attribution to specific sources, with some confidence. Today, ever greater detailed information on each strain is being acquired with the use of whole-genome sequences. Unfortunately, it is a consequence of the wealth of such data, and the sophistication of the analytical techniques required, that only experts can now interpret the information obtained.

Overall, *Campylobacter* is widely considered a foodborne pathogen, with the handling and consumption of poultry meat implicated as the major source. As a consequence, risk communication and management have focused on the poultry industry. However, there is growing evidence that alternative nonmeat-foodborne transmission routes, such as the contact with animals directly, or indirectly through the environment (including vegetables) they contaminate, have a major role in campylobacteriosis.

REFERENCES

Abley, M.J., Wittum, T.E., Funk, J.A., Gebreyes, W.A., 2012. Antimicrobial susceptibility, pulsed-field gel electrophoresis, and multi-locus sequence typing of *Campylobacter coli* in swine before, during, and after the slaughter process. Foodborne Pathog. Dis. 9, 506–512.

Acke, E., McGill, K., Golden, O., Jones, B.R., Fanning, S., Whyte, P., 2009. Prevalence of thermophilic *Campylobacter* species in household cats and dogs in Ireland. Vet. Rec. 164, 44–47.

Allen, V.M., Ridley, A.M., Harris, J.A., Newell, D.G., Powell, L., 2011. Influence of production system on the rate of onset of *Campylobacter* colonization in chicken flocks reared extensively in the United Kingdom. Br. Poult. Sci. 52, 30–39.

Ansari-Lari, M., Hosseinzadeh, S., Shekarforoush, S.S., Abdollahi, M., Berizi, E., 2011. Prevalence and risk factors associated with *Campylobacter* infections in broiler flocks in Shiraz, southern Iran. Int. J. Food Microbiol. 144, 475–479.

Arnone, R.D., Walling, J.P., 2007. Waterborne pathogens in urban watersheds. J. Water Health 5, 149–162.

Besser, T.E., Lejeune, J.T., Rice, D.H., Berg, J., Stilborn, R.P., Kaya, K., Bae, W., Hancock, D.D., 2005. Increasing prevalence of *Campylobacter jejuni* in feedlot cattle through the feeding period. Appl. Environ. Microbiol. 71, 5752–5758.

Bianchini, V., Borella, L., Benedetti, V., Parisi, A., Miccolupo, A., Santoro, E., Recordati, C., Luini, M., 2014. Prevalence in bulk tank milk and epidemiology of *Campylobacter jejuni* in dairy herds in Northern Italy. Appl. Environ. Microbiol. 80, 1832–1837.

Bouwknegt, M., van Pelt, W., Kubbinga, M.E., Weda, M., Havelaar, A.H., 2014. Potential association between the recent increase in campylobacteriosis incidence in the Netherlands and proton-pump inhibitor use—an ecological study. Euro Surveill. 19 (32), 20873.

Carbonero, A., Torralbo, A., Borge, C., García-Bocanegra, I., Arenas, A., Perea, A., 2012. *Campylobacter* spp., *C. jejuni* and *C. upsaliensis* infection-associated factors in healthy and ill dogs from clinics in Cordoba, Spain. Screening tests for antimicrobial susceptibility. Comp. Immunol. Microbiol. Infect. Dis. 35, 505–512.

Carrique-Mas, J.J., Bryant, J.E., Cuong, N.V., Hoang, N.V.M., Campbell, J., Hoang, N.V., Dung, T.T.N., Duy, D.T., Hoa, N.T., Thompson, C., Hien, V.V., Phat, V.V., Farrar, J., Baker, S., 2014. An epidemiological investigation of *Campylobacter* in pig and poultry farms in the Mekong Delta of Vietnam. Epidemiol. Infect. 142, 1425–1436.

Cawthraw, S.A., Newell, D.G., 2010. Investigation of the presence and protective effects of maternal antibodies against *Campylobacter jejuni* in chickens. Avian Dis. 54, 86–93.

Ceuppens, S., Johannessen, G.S., Allende, A., Tondo, E.C., El-Tahan, F., Sampers, I., Jacxsens, L., Uyttendaele, M., 2015. Risk factors for *Salmonella*, shiga toxin-producing *Escherichia coli* and *Campylobacter* occurrence in primary production of leafy greens and strawberries. Int. J. Environ. Res. Public Health 12, 9809–9831.

Chaban, B., Ngeleka, M., Hill, J.E., 2010. Detection and quantification of 14 *Campylobacter* species in pet dogs reveals an increase in species richness in feces of diarrheic animals. BMC Microbiol. 10, 73.

Clark, C.G., Price, L., Ahmed, R., Woodward, D.L., Melito, P.L., Rodgers, F.G., Jamieson, F., Ciebin, B., Li, A., Ellis, A., 2003. Characterization of waterborne outbreak-associated *Campylobacter jejuni*, Walkerton, Ontario. Emerg. Infect. Dis. 9, 1232–1241.

Cody, A.J., McCarthy, N.D., Bray, J.E., Wimalarathna, H.M.L., Colles, F.M., Jansen van Rensburg, M.J., Dingle, K.E., Waldenström, J., Maiden, M.C.J., 2015. Wild bird-associated *Campylobacter jejuni* isolates are a consistent source of human disease, in Oxfordshire, United Kingdom. Environ. Microbiol. Rep. 7, 782–788.

Colles, F.M., Ali, J.S., Sheppard, S.K., McCarthy, N.D., Maiden, M.C.J., 2011. *Campylobacter* populations in wild and domesticated Mallard ducks (*Anas platyrhynchos*). Environ. Microbiol. Rep. 3, 574–580.

Conan, A., Goutard, F.L., Sorn, S., Vong, S., 2012. Biosecurity measures for backyard poultry in developing countries: a systematic review. BMC Vet. Res. 8, 240.

Danis, K., Di Renzi, M., O'Neill, W., Smyth, B., McKeown, P., Foley, B., Tohani, V., Devine, M., 2009. Risk factors for sporadic *Campylobacter* infection: an all-Ireland case-control study. Euro Surveill. 14 (7), 19123.

de Carvalho, A.F., da Silva, D.M., Azevedo, S.S., Piatti, R.M., Genovez, M.E., Scarcelli, E., 2013. Detection of CDT toxin genes in *Campylobacter* spp. strains isolated from broiler carcasses and vegetables in São Paulo, Brazil. Braz. J. Microbiol. 44, 693–699.

Dearlove, B.L., Cody, A.J., Pascoe, B., Méric, G., Wilson, D.J., Sheppard, S.K., 2016. Rapid host switching in generalist *Campylobacter* strains erodes the signal for tracing human infections. ISME J. 10 (3), 721–729.

Dingle, K.E., Colles, F.M., Wareing, D.R., Ure, R., Fox, A.J., Bolton, F.E., Bootsma, H.J., Willems, R.J., Urwin, R., Maiden, M.C., 2001. Multilocus sequence typing system for *Campylobacter jejuni*. J. Clin. Microbiol. 39, 14–23.

Domingues, A.R., Pires, S.M., Halasa, T., Hald, T., 2012. Source attribution of human campylobacteriosis using a meta-analysis of case-control studies of sporadic infections. Epidemiol. Infect. 140, 970–981.

Doorduyn, Y., Van Den Brandhof, W.E., Van Duynhoven, Y.T.H.P., Breukink, B.J., Wagenaar, J.A., Van Pelt, W., 2010. Risk factors for indigenous *Campylobacter jejuni* and *Campylobacter coli* infections in The Netherlands: a case-control study. Epidemiol. Infect. 138, 1391–1404.

EFSA (European Food Safety Authority), 2010a. Analysis of the baseline survey on the prevalence of *Campylobacter* in broiler batches, and of *Campylobacter* and *Salmonella* on broiler carcasses in the EU, 2008; Part A: *Campylobacter* and *Salmonella* prevalence estimates. EFSA J. 8 (1503), 99.

EFSA (European Food Safety Authority), 2010b. Analysis of the baseline survey on the prevalence of *Campylobacter* in broiler batches, and of *Campylobacter* and *Salmonella* on broiler carcasses, in the EU, 2008; Part B: Analysis of factors associated with *Campylobacter* colonisation of broiler batches and with *Campylobacter* contamination of broiler carcasses; and investigation of the culture method diagnostic characteristics used to analyse broiler carcass samples. EFSA J. 8 (8), 1522, [132 pp.].

EFSA Panel on Biological Hazards (BIOHAZ), 2010c. Scientific opinion on quantification of the risk posed by broiler meat to human campylobacteriosis in the EU. EFSA J. 8 (1), 1437, [89 pp.].

EFSA (European Food Safety Authority), ECDC (European Centre for Disease Prevention and Control), 2015. The European Union summary report on trends and sources of zoonoses, zoonotic agents and food-borne outbreaks in 2014. EFSA J. 13 (12), 4329, [191 pp.].

Ellis-Iversen, J., Pritchard, G.C., Wooldridge, M., Nielen, M., 2009. Risk factors for *Campylobacter jejuni* and *Campylobacter coli* in young cattle on English and Welsh farms. Prev. Vet. Med. 88, 42–48.

Elson, R., Burgess, F., Little, C.L., Mitchell, R.T., 2004. Microbiological examination of ready-to-eat cold sliced meats and pate from catering and retail premises in the UK. J. Appl. Microbiol. 96, 499–509.

Evers, E.G., Blaak, H., Hamidjaja, R.A., de Jonge, R., Schets, F.M., 2016. A QMRA for the transmission of ESBL-producing *Escherichia coli* and *Campylobacter* from poultry farms to humans through flies. Risk Anal. 36, 215–227.

Fernandes, A.M., Balasegaram, S., Willis, C., Wimalarathna, H.M.L., Maiden, M.C., McCarthy, N.D., 2015. Partial failure of milk pasteurization as a risk for the transmission of *Campylobacter* from cattle to humans. Clin. Infect. Dis. 61, 903–909.

Förster, M., Sievert, K., Messler, S., Klimpel, S., Pfeffer, K., 2009. Comprehensive study on the occurrence and distribution of pathogenic microorganisms carried by synanthropic flies caught at different rural locations in Germany. J. Med. Entomol. 46, 1164–1166.

French, N., Barrigas, M., Brown, P., Ribiero, P., Williams, N., Leatherbarrow, H., Birtles, R., Bolton, E., Fearnhead, P., Fox, A., 2005. Spatial epidemiology and natural population structure of *Campylobacter jejuni* colonizing a farmland ecosystem. Environ. Microbiol. 7, 1116–1126.

French, N.P., Midwinter, A., Holland, B., Collins-Emerson, J., Pattison, R., Colles, F., Carter, P., 2009. Molecular epidemiology of *Campylobacter jejuni* isolates from wild-bird fecal material in children's playgrounds. Appl. Environ. Microbiol. 75, 779–783.

Friedman, C.R., Hoekstra, R.M., Samuel, M., Marcus, R., Bender, J., Shiferaw, B., Reddy, S., Ahuja, S.D., Helfrick, D.L., Hardnett, F., Carter, M., Anderson, B., Tauxe, R.V., Emerging Infections Program FoodNet Working Group, 2004. Risk factors for sporadic *Campylobacter* infection in the United States: a case-control study in FoodNet sites. Clin. Infect. Dis. 38 (Suppl. 3), S285–S296.

Friesema, I.H.M., Havelaar, A.H., Westra, P.P., Wagenaar, J.A., van Pelt, W., 2012. Poultry culling and campylobacteriosis reduction among humans, the Netherlands. Emerg. Infect. Dis. 18, 466–468.

Fullerton, K.E., Ingram, L.A., Jones, T.F., Anderson, B.J., McCarthy, P.V., Hurd, S., Shiferaw, B., Vugia, D., Haubert, N., Hayes, T., Wedel, S., Scallan, E., Henao, O., Angulo, F.J., 2007. Sporadic *Campylobacter* infection in infants: a population-based surveillance case-control study. Pediatr. Infect. Dis. J. 26, 19–24.

Gallay, A., Bousquet, V., Siret, V., Prouzet-Mauléon, V., Valk, H., de Vaillant, V., Simon, F., Le Strat, Y., Mégraud, F., Desenclos, J.-C., 2008. Risk factors for acquiring sporadic *Campylobacter* infection in France: results from a national case-control study. J. Infect. Dis. 197, 1477–1484.

Gardner, T.J., Fitzgerald, C., Xavier, C., Klein, R., Pruckler, J., Stroika, S., McLaughlin, J.B., 2011. Outbreak of campylobacteriosis associated with consumption of raw peas. Clin. Infect. Dis. 53, 26–32.

Gilbert, M., Conchedda, G., Van Boeckel, T.P., Cinardi, G., Linard, C., Nicolas, G., Thanapongtharm, W., D'Aietti, L., Wint, W., Newman, S.H., Robinson, T.P., 2015. Income disparities and the global distribution of intensively farmed chicken and pigs. PLoS One 10, e0133381.

Griekspoor, P., Colles, F.M., McCarthy, N.D., Hansbro, P.M., Ashhurst-Smith, C., Olsen, B., Hasselquist, D., Maiden, M.C.J., Waldenström, J., 2013. Marked host specificity and lack of phylogeographic population structure of *Campylobacter jejuni* in wild birds. Mol. Ecol. 22, 1463–1472.

Hakkinen, M., Hänninen, M.-L., 2009. Shedding of *Campylobacter* spp. in Finnish cattle on dairy farms. J. Appl. Microbiol. 107, 898–905.

Hakkinen, M., Heiska, H., Hänninen, M.-L., 2007. Prevalence of *Campylobacter* spp. in cattle in Finland and antimicrobial susceptibilities of bovine *Campylobacter jejuni* strains. Appl. Environ. Microbiol. 73, 3232–3238.

Harrison, W.A., Griffith, C.J., Tennant, D., Peters, A.C., 2001. Incidence of *Campylobacter* and *Salmonella* isolated from retail chicken and associated packaging in South Wales. Lett. Appl. Microbiol. 33, 450–454.

Havelaar, A.H., van Pelt, W., Ang, C.W., Wagenaar, J.A., van Putten, J.P.M., Gross, U., Newell, D.G., 2009. Immunity to *Campylobacter*: its role in risk assessment and epidemiology. Crit. Rev. Microbiol. 35, 1–22.

Hill, B., Smythe, B., Lindsay, D., Shepherd, J., 2012. Microbiology of raw milk in New Zealand. Int. J. Food Microbiol. 157, 305–308.

Hoar, B.R., Atwill, E.R., Elmi, C., Farver, T.B., 2001. An examination of risk factors associated with beef cattle shedding pathogens of potential zoonotic concern. Epidemiol. Infect. 127, 147–155.

Huang, J., Zong, Q., Zhao, F., Zhu, J., Jiao, X.-a., 2016. Quantitative surveys of *Salmonella* and *Campylobacter* on retail raw chicken in Yangzhou, China. Food Control 59, 68–73.

Jacobs-Reitsma, W., Lyhs, U.W.J., 2008. *Campylobacter* in the food supply. In: Nachamkin, I., Szymanski, C.M., Blaser, M.J. (Eds.), Campylobacter. third ed. ASM Press, Washington, DC, pp. 627–644.

Johnsen, G., Zimmerman, K., Lindstedt, B.-A., Vardund, T., Herikstad, H., Kapperud, G., 2006. Intestinal carriage of *Campylobacter jejuni* and *Campylobacter coli* among cattle from south-western Norway and comparative genotyping of bovine and human isolates by amplified-fragment length polymorphism. Acta Vet. Scand. 48, 4.

Jonas, R., Kittl, S., Overesch, G., Kuhnert, P., 2015. Genotypes and antibiotic resistance of bovine *Campylobacter* and their contribution to human campylobacteriosis. Epidemiol. Infect. 143, 2373–2380.

Jones, K., 2001. Campylobacters in water, sewage and the environment. Symp. Ser. Soc. Appl. Microbiol. 30, 68S–79S.

Jones, K., Howard, S., Wallace, J.S., 1999. Intermittent shedding of thermophilic campylobacters by sheep at pasture. J. Appl. Microbiol. 86, 531–536.

Kalupahana, R.S., Kottawatta, K.S., Kanankege, K.S., van Bergen, M.A., Abeynayake, P., Wagenaar, J.A., 2013. Colonization of *Campylobacter* spp. in broiler chickens and laying hens reared in tropical climates with low-biosecurity housing. Appl. Environ. Microbiol. 79, 393–395.

Kapperud, G., Espeland, G., Wahl, E., Walde, A., Herikstad, H., Gustavsen, S., Tveit, I., Natås, O., Bevanger, L., Digranes, A., 2003. Factors associated with increased and decreased risk of *Campylobacter* infection: a prospective case-control study in Norway. Am. J. Epidemiol. 158, 234–242.

Kärenlampi, R., Hänninen, M.-L., 2004. Survival of *Campylobacter jejuni* on various fresh produce. Int. J. Food Microbiol. 97, 187–195.

Khalid, M.I., Tang, J.Y.H., Baharuddin, N.H., Rahman, N.S., Rahimi, N.F., Radu, S., 2015. Prevalence, antibiogram, and cdt genes of toxigenic *Campylobacter jejuni* in salad style vegetables (ulam) at farms and retail outlets in Terengganu. J. Food Prot. 78, 65–71.

Kolstoe, E.M., Iversen, T., Østensvik, Ø., Abdelghani, A., Secic, I., Nesbakken, T., 2015. Specific pathogen-free pig herds also free from *Campylobacter*? Zoonoses Public Health 62, 125–130.

Korsak, D., Maćkiw, E., Rożynek, E., Żyłowska, M., 2015. Prevalence of *Campylobacter* spp. in retail chicken, turkey, pork, and beef meat in Poland between 2009 and 2013. J. Food Prot. 78, 1024–1028.

Lazou, T., Houf, K., Soultos, N., Dovas, C., Iossifidou, E., 2014. *Campylobacter* in small ruminants at slaughter: prevalence, pulsotypes and antibiotic resistance. Int. J. Food Microbiol. 173, 54–61.

Little, C.L., Richardson, J.F., Owen, R.J., de Pinna, E., Threlfall, E.J., 2008. *Campylobacter* and *Salmonella* in raw red meats in the United Kingdom: prevalence, characterization and antimicrobial resistance pattern, 2003–2005. Food Microbiol. 25, 538–543.

Losio, M.N., Pavoni, E., Bilei, S., Bertasi, B., Bove, D., Capuano, F., Farneti, S., Blasi, G., Comin, D., Cardamone, C., Decastelli, L., Delibato, E., De Santis, P., Di Pasquale, S., Gattuso, A., Goffredo, E., Fadda, A., Pisanu, M., De Medici, D., 2015. Microbiological survey of raw and ready-to-eat leafy green vegetables marketed in Italy. Int. J. Food Microbiol. 210, 88–91.

Luber, P., Brynestad, S., Topsch, D., Scherer, K., Bartelt, E., 2006. Quantification of *Campylobacter* species cross-contamination during handling of contaminated fresh chicken parts in kitchens. Appl. Environ. Microbiol. 72, 66–70.

Ma, L., Wang, Y., Shen, J., Zhang, Q., Wu, C., 2014. Tracking *Campylobacter* contamination along a broiler chicken production chain from the farm level to retail in China. Int. J. Food Microbiol. 181, 77–84.

Maćkiw, E., Rzewuska, K., Stoś, K., Jarosz, M., Korsak, D., 2011. Occurrence of *Campylobacter* spp. in poultry and poultry products for sale on the Polish retail market. J. Food Prot. 74, 986–989.

Marks, S.L., Rankin, S.C., Byrne, B.A., Weese, J.S., 2011. Enteropathogenic bacteria in dogs and cats: diagnosis, epidemiology, treatment, and control. J. Vet. Intern. Med. 25, 1195–1208.

Mattheus, W., Botteldoorn, N., Heylen, K., Pochet, B., Dierick, K., 2012. Trend analysis of antimicrobial resistance in *Campylobacter jejuni* and *Campylobacter coli* isolated from Belgian pork and poultry meat products using surveillance data of 2004–2009. Foodborne Pathog. Dis. 9, 465–472.

McCarthy, N.D., Colles, F.M., Dingle, K.E., Bagnall, M.C., Manning, G., Maiden, M.C., Falush, D., 2007. Host-associated genetic import in *Campylobacter jejuni*. Emerg. Infect. Dis. 13, 267–272.

Meldrum, R.J., Ribeiro, C.D., 2003. *Campylobacter* in ready-to-eat foods: the result of a 15-month survey. J. Food Prot. 66, 2135–2137.

Merialdi, G., Giacometti, F., Bardasi, L., Stancampiano, L., Taddei, R., Serratore, P., Serraino, A., 2015. Fecal shedding of thermophilic *Campylobacter* in a dairy herd producing raw milk for direct human consumption. J. Food Prot. 78, 579–584.

Milnes, A.S., Stewart, I., Clifton-Hadley, F.A., Davies, R.H., Newell, D.G., Sayers, A.R., Cheasty, T., Cassar, C., Ridley, A., Cook, A.J.C., Evans, S.J., Teale, C.J., Smith, R.P., McNally, A., Toszeghy, M., Futter, R., Kay, A., Paiba, G.A., 2008. Intestinal carriage of verocytotoxigenic *Escherichia coli* O157, *Salmonella*, thermophilic *Campylobacter* and *Yersinia enterocolitica*, in cattle, sheep and pigs at slaughter in Great Britain during 2003. Epidemiol. Infect. 136, 739–751.

Moore, J.E., Wilson, T.S., Wareing, D.R.A., Humphrey, T.J., Murphy, P.G., 2002. Prevalence of thermophilic *Campylobacter* spp. in ready-to-eat foods and raw poultry in Northern Ireland. J. Food Prot. 65, 1326–1328.

Moriarty, E.M., Downing, M., Bellamy, J., Gilpin, B.J., 2015. Concentrations of faecal coliforms, *Escherichia coli*, enterococci and *Campylobacter* spp. in equine faeces. NZ Vet. J. 63, 104–109.

Mossong, J., Mughini-Gras, L., Penny, C., Devaux, A., Olinger, C., Losch, S., Cauchie, H.-M., van Pelt, W., Ragimbeau, C., 2016. Human campylobacteriosis in Luxembourg, 2010–2013: a case-control study combined with multilocus sequence typing for source attribution and risk factor analysis. Sci. Rep. 6, 20939.

Muehlherr, J.E., Zweifel, C., Corti, S., Blanco, J.E., Stephan, R., 2003. Microbiological quality of raw goat's and ewe's bulk-tank milk in Switzerland. J. Dairy Sci. 86, 3849–3856.

Muellner, P., Marshall, J.C., Spencer, S.E., Noble, A.D., Shadbolt, T., Collins-Emerson, J.M., Midwinter, A.C., Carter, P.E., Pirie, R., Wilson, D.J., Campbell, D.M., Stevenson, M.A., French, N.P., 2011. Utilizing a combination of molecular and spatial tools to assess the effect of a public health intervention. Prev. Vet. Med. 102, 242–253.

Mughini Gras, L., Smid, J.H., Wagenaar, J.A., de Boer, A.G., Havelaar, A.H., Friesema, I.H.M., French, N.P., Busani, L., van Pelt, W., 2012. Risk factors for campylobacteriosis of chicken, ruminant, and environmental origin: a combined case-control and source attribution analysis. PLoS One 7, e42599.

Mughini Gras, L., Smid, J.H., Wagenaar, J.A., Koene, M.G.J., Havelaar, A.H., Friesema, I.H.M., French, N.P., Flemming, C., Galson, J.D., Graziani, C., Busani, L., van Pelt, W., 2013. Increased risk for *Campylobacter jejuni* and *C. coli* infection of pet origin in dog owners and evidence for genetic association between strains causing infection in humans and their pets. Epidemiol. Infect. 141, 2526–2535.

Mughini-Gras, L., Smid, J.H., Wagenaar, J.A., De Boer, A., Havelaar, A.H., Friesema, I.H.M., French, N.P., Graziani, C., Busani, L., van Pelt, W., 2014. Campylobacteriosis in returning travellers and potential secondary transmission of exotic strains. Epidemiol. Infect. 142, 1277–1288.

Mullner, P., Jones, G., Noble, A., Spencer, S.E.F., Hathaway, S., French, N.P., 2009a. Source attribution of food-borne zoonoses in New Zealand: a modified Hald model. Risk Anal. 29, 970–984.

Mullner, P., Spencer, S.E.F., Wilson, D.J., Jones, G., Noble, A.D., Midwinter, A.C., Collins-Emerson, J.M., Carter, P., Hathaway, S., French, N.P., 2009b. Assigning the source of human campylobacteriosis in New Zealand: a comparative genetic and epidemiological approach. Infect. Genet. Evol. 9, 1311–1319.

Nathues, C., Grüning, P., Fruth, A., Verspohl, J., Blaha, T., Kreienbrock, L., Merle, R., 2013. *Campylobacter* spp., *Yersinia enterocolitica*, and *Salmonella enterica* and their simultaneous occurrence in German fattening pig herds and their environment. J. Food Prot. 76, 1704–1711.

Neal, K.R., Slack, R.C., 1997. Diabetes mellitus, anti-secretory drugs and other risk factors for *Campylobacter* gastro-enteritis in adults: a case-control study. Epidemiol. Infect. 119, 307–311.

Newell, D.G., Elvers, K.T., Dopfer, D., Hansson, I., Jones, P., James, S., Gittins, J., Stern, N.J., Davies, R., Connerton, I., Pearson, D., Salvat, G.S., Allen, V.M., 2011. Biosecurity-based interventions and strategies to reduce *Campylobacter* spp. on poultry farms. Appl. Environ. Microbiol. 77, 8605–8614.

Nielsen, E.M., 2002. Occurrence and strain diversity of thermophilic campylobacters in cattle of different age groups in dairy herds. Lett. Appl. Microbiol. 35, 85–89.

Noormohamed, A., Fakhr, M.K., 2013. A higher prevalence rate of *Campylobacter* in retail beef livers compared to other beef and pork meat cuts. Int. J. Environ. Res. Public Health 10, 2058–2068.

Olofsson, J., Berglund, P.G., Olsen, B., Ellström, P., Axelsson-Olsson, D., 2015. The abundant free-living amoeba, *Acanthamoeba polyphaga*, increases the survival of *Campylobacter jejuni* in milk and orange juice. Infect. Ecol. Epidemiol. 5, 28675.

Orr, K.E., Lightfoot, N.F., Sisson, P.R., Harkis, B.A., Tweddle, J.L., Boyd, P., Carroll, A., Jackson, C.J., Wareing, D.R., Freeman, R., 1995. Direct milk excretion of *Campylobacter jejuni* in a dairy cow causing cases of human enteritis. Epidemiol. Infect. 114, 15–24.

Parkar, S.F., Sachdev, D., deSouza, N., Kamble, A., Suresh, G., Munot, H., 2013. Prevalence, seasonality and antibiotic susceptibility of thermophilic campylobacters in caeca and carcasses of poultry birds in the live-bird market. Afr. J. Microbiol. Res. 7 (21), 2442–2453.

Pascoe, B., Méric, G., Murray, S., Yahara, K., Mageiros, L., Bowen, R., Jones, N.H., Jeeves, R.E., Lappin-Scott, H.M., Asakura, H., Sheppard, S.K., 2015. Enhanced biofilm formation and multi-host transmission evolve from divergent genetic backgrounds in *Campylobacter jejuni*. Environ. Microbiol. 17, 4779–4789.

Pires, S.M., Evers, E.G., van Pelt, W., Ayers, T., Scallan, E., Angulo, F.J., Havelaar, A., Hald, T., Med-Vet-Net Workpackage 28 Working Group, 2009. Attributing the human disease burden of foodborne infections to specific sources. Foodborne Pathog. Dis. 6, 417–424.

Pires, S.M., Vigre, H., Makela, P., Hald, T., 2010. Using outbreak data for source attribution of human salmonellosis and campylobacteriosis in Europe. Foodborne Pathog. Dis. 7, 1351–1361.

Platts-Mills, J.A., Babji, S., Bodhidatta, L., Gratz, J., Haque, R., Havt, A., McCormick, B.J., McGrath, M., Olortegui, M.P., Samie, A., Shakoor, S., Mondal, D., Lima, I.F., Hariraju, D., Rayamajhi, B.B., Qureshi, S., Kabir, F., Yori, P.P., Mufamadi, B., Amour, C., Carreon, J.D., Richard, S.A., Lang, D., Bessong, P., Mduma, E., Ahmed, T., Lima, A.A., Mason, C.J., Zaidi, A.K., Bhutta, Z.A., Kosek, M., Guerrant, R.L., Gottlieb, M., Miller, M., Kang, G., Houpt, E.R., MAL-ED Network Investigators, 2015. Pathogen-specific burdens of community diarrhoea in developing countries: a multisite birth cohort study (MAL-ED). Lancet Glob. Health 3, e564–e575.

Psifidi, A., Fife, M., Howell, J., Matika, O., van Diemen, P.M., Kuo, R., Smith, J., Hocking, P.M., Salmon, N., Jones, M.A., Hume, D.A., Banos, G., Stevens, M.P., Kaiser, P., 2016. The genomic architecture of resistance to *Campylobacter jejuni* intestinal colonisation in chickens. BMC Genomics 17, 293.

Quintana-Hayashi, M.P., Thakur, S., 2012. Longitudinal study of the persistence of antimicrobial-resistant *Campylobacter* strains in distinct swine production systems on farms, at slaughter, and in the environment. Appl. Environ. Microbiol. 78, 2698–2705.

Rodrigues, L.C., Wheeler, J.G., Sethi, D., Wall, P.G., Cumberland, P., Tompkins, D.S., Hudson, M.J., Roberts, J.A., Roderick, P.J., 2000. The study of infectious intestinal disease in England: risk factors for cases of infectious intestinal disease with *Campylobacter jejuni* infection. Epidemiol. Infect. 127, 185–193.

Rotariu, O., Dallas, J.F., Ogden, I.D., MacRae, M., Sheppard, S.K., Maiden, M.C.J., Gormley, F.J., Forbes, K.J., Strachan, N.J.C., 2009. Spatiotemporal homogeneity of *Campylobacter* subtypes from cattle and sheep across northeastern and southwestern Scotland. Appl. Environ. Microbiol. 75, 6275–6281.

Roug, A., Byrne, B.A., Conrad, P.A., Miller, W.A., 2013. Zoonotic fecal pathogens and antimicrobial resistance in county fair animals. Comp. Immunol. Microbiol. Infect. Dis. 36, 303–308.

Ruusunen, M., Salonen, M., Pulkkinen, H., Huuskonen, M., Hellström, S., Revez, J., Hänninen, M.-L., Fredriksson-Ahomaa, M., Lindström, M., 2013. Pathogenic bacteria in Finnish bulk tank milk. Foodborne Pathog. Dis. 10, 99–106.

Salihu, M., Junaidu, A., Magaji, A., Abubakar, M., Adamu, A., Yakubu, A., 2009. Prevalence of *Campylobacter* in poultry meat in Sokoto, Northwestern Nigeria. J. Public Health Epidemiol. 1, 041–045.

Sammarco, M.L., Ripabelli, G., Fanelli, I., Grasso, G.M., Tamburro, M., 2010. Prevalence and biomolecular characterization of *Campylobacter* spp. isolated from retail meat. J. Food Prot. 73, 720–728.

Sasaki, Y., Murakami, M., Haruna, M., Maruyama, N., Mori, T., Ito, K., Yamada, Y., 2013. Prevalence and characterization of foodborne pathogens in dairy cattle in the eastern part of Japan. J. Vet. Med. Sci. 75, 543–546.

Sato, K., Bartlett, P.C., Kaneene, J.B., Downes, F.P., 2004. Comparison of prevalence and antimicrobial susceptibilities of *Campylobacter* spp. isolates from organic and conventional dairy herds in Wisconsin. Appl. Environ. Microbiol. 70, 1442–1447.

Sawabe, T., Suda, W., Ohshima, K., Hattori, M., Sawabe, T., 2015. First microbiota assessments of children's paddling pool waters evaluated using 16S rRNA gene-based metagenome analysis. J. Infect. Public Health 9, 362–365.

Sears, A., Baker, M.G., Wilson, N., Marshall, J., Muellner, P., Campbell, D.M., Lake, R.J., French, N.P., 2011. Marked campylobacteriosis decline after interventions aimed at poultry. NZ Emerg. Infect. Dis. 17, 1007–1015.

Sheppard, S.K., Dallas, J.F., Strachan, N.J.C., MacRae, M., McCarthy, N.D., Wilson, D.J., Gormley, F.J., Falush, D., Ogden, I.D., Maiden, M.C.J., Forbes, K.J., 2009. *Campylobacter* genotyping to determine the source of human infection. Clin. Infect. Dis. 48, 1072–1078.

Skirrow, M.B., 1977. *Campylobacter* enteritis: a "new" disease. Br. Med. J. 2, 9–11.

Sopwith, W., Birtles, A., Matthews, M., Fox, A., Gee, S., Painter, M., Regan, M., Syed, Q., Bolton, E., 2008. Identification of potential environmentally adapted *Campylobacter jejuni* strain, United Kingdom. Emerg. Infect. Dis. 14, 1769–1773.

Stafford, R.J., Schluter, P., Kirk, M., Wilson, A., Unicomb, L., Ashbolt, R., Gregory, J., OzFoodNet Working Group, 2007. A multi-centre prospective case-control study of *Campylobacter* infection in persons aged 5 years and older in Australia. Epidemiol. Infect. 135, 978–988.

Stanley, K.N., Wallace, J.S., Currie, J.E., Diggle, P.J., Jones, K., 1998a. The seasonal variation of thermophilic campylobacters in beef cattle, dairy cattle and calves. J. Appl. Microbiol. 85, 472–480.

Stanley, K.N., Wallace, J.S., Currie, J.E., Diggle, P.J., Jones, K., 1998b. Seasonal variation of thermophilic campylobacters in lambs at slaughter. J. Appl. Microbiol. 84, 1111–1116.

Stavisky, J., Radford, A.D., Gaskell, R., Dawson, S., German, A., Parsons, B., Clegg, S., Newman, J., Pinchbeck, G., 2011. A case-control study of pathogen and lifestyle risk factors for diarrhoea in dogs. Prev. Vet. Med. 99, 185–192.

Stegeman, A., Bouma, A., Elbers, A.R.W., de Jong, M.C.M., Nodelijk, G., de Klerk, F., Koch, G., van Boven, M., 2004. Avian influenza A virus (H7N7) epidemic in The Netherlands in 2003: course of the epidemic and effectiveness of control measures. J. Infect. Dis. 190, 2088–2095.

Stern, N.J., Hiett, K.L., Alfredsson, G.A., Kristinsson, K.G., Reiersen, J., Hardardottir, H., Briem, H., Gunnarsson, E., Georgsson, F., Lowman, R., Berndtson, E., Lammerding, A.M., Paoli, G.M., Musgrove, M.T., 2003. *Campylobacter* spp. in Icelandic poultry operations human disease. Epidemiol. Infect. 130, 23–32.

Strachan, N.J.C., Gormley, F.J., Rotariu, O., Ogden, I.D., Miller, G., Dunn, G.M., Sheppard, S.K., Dallas, J.F., Reid, T.M.S., Howie, H., Maiden, M.C.J., Forbes, K.J., 2009. Attribution of *Campylobacter* infections in northeast Scotland to specific sources by use of multilocus sequence typing. J. Infect. Dis. 199, 1205–1208.

Studahl, A., Andersson, Y., 2000. Risk factors for indigenous *Campylobacter* infection: a Swedish case-control study. Epidemiol. Infect. 125, 269–275.

Swift, L., Hunter, P.R., 2004. What do negative associations between potential risk factors and illness in analytical epidemiological studies of infectious disease really mean? Eur. J. Epidemiol. 19, 219–223.

Tam, C.C., Higgins, C.D., Neal, K.R., Rodrigues, L.C., Millership, S.E., O'Brien, S.J., *Campylobacter* Case-Control Study Group, 2009. Chicken consumption and use of acid-suppressing medications as risk factors for *Campylobacter* enteritis, England. Emerg. Infect. Dis. 15, 1402–1408.

Trigui, H., Paquet, V.E., Charette, S.J., Faucher, S.P., 2016. Packaging of *Campylobacter jejuni* into multilamellar bodies by the ciliate *Tetrahymena pyriformis*. Appl. Environ. Microbiol. 82, 2783–2790.

Vellinga, A., Van Loock, F., 2002. The dioxin crisis as experiment to determine poultry-related *Campylobacter* enteritis. Emerg. Infect. Dis. 8, 19–22.

Verhoeff-Bakkenes, L., Jansen, H.A.P.M., in 't Veld, P.H., Beumer, R.R., Zwietering, M.H., van Leusden, F.M., 2011. Consumption of raw vegetables and fruits: a risk factor for *Campylobacter* infections. Int. J. Food Microbiol. 144, 406–412.

Wagenaar, J.A., French, N.P., Havelaar, A.H., 2013. Preventing *Campylobacter* at the source: why is it so difficult? Clin. Infect. Dis. 57, 1600–1606.

Wagenaar, J.A., Jacobs-Reitsma, W., Hofshagen, M., Newell, D.G., 2008. Poultry colonisation with *Campylobacter* and its control at the primary production level. In: Nachamkin, I., Szymanski, C.M., Blaser, M.J. (Eds.), Campylobacter. third ed. ASM Press, Washington, DC, pp. 667–678.

Wesley, I.V., Wells, S.J., Harmon, K.M., Green, A., Schroeder-Tucker, L., Glover, M., Siddique, I., 2000. Fecal shedding of *Campylobacter* and *Arcobacter* spp. in dairy cattle. Appl. Environ. Microbiol. 66, 1994–2000.

Whiley, H., van den Akker, B., Giglio, S., Bentham, R., 2013. The role of environmental reservoirs in human campylobacteriosis. Int. J. Environ. Res. Public Health 10, 5886–5907.

Whyte, P., McGill, K., Cowley, D., Madden, R.H., Moran, L., Scates, P., Carroll, C., O'Leary, A., Fanning, S., Collins, J.D., McNamara, E., Moore, J.E., Cormican, M., 2004. Occurrence of *Campylobacter* in retail foods in Ireland. Int. J. Food Microbiol. 95, 111–118.

Wieland, B., Regula, G., Danuser, J., Wittwer, M., Burnens, A.P., Wassenaar, T.M., Stärk, K.D.C., 2005. *Campylobacter* spp. in dogs and cats in Switzerland: risk factor analysis and molecular characterization with AFLP. J. Vet. Med. B 52, 183–189.

Wijnands, L.M., Delfgou-van Asch, E.H.M., Beerepoot-Mensink, M.E., van der Meij-Florijn, A., Fitz-James, I., van Leusden, F.M., Pielaat, A., 2014. Prevalence and concentration of bacterial pathogens in raw produce and minimally processed packaged salads produced in and for the Netherlands. J. Food Prot. 77, 388–394.

Wilson, D.J., Gabriel, E., Leatherbarrow, A.J.H., Cheesbrough, J., Gee, S., Bolton, E., Fox, A., Fearnhead, P., Hart, C.A., Diggle, P.J., 2008. Tracing the source of campylobacteriosis. PLoS Genet. 4, e1000203.

Wong, T.L., Hollis, L., Cornelius, A., Nicol, C., Cook, R., Hudson, J.A., 2007. Prevalence, numbers, and subtypes of *Campylobacter jejuni* and *Campylobacter coli* in uncooked retail meat samples. J. Food Prot. 70, 566–573.

Zhao, S., Young, S.R., Tong, E., Abbott, J.W., Womack, N., Friedman, S.L., McDermott, P.F., 2010. Antimicrobial resistance of *Campylobacter* isolates from retail meat in the United States between 2002 and 2007. Appl. Environ. Microbiol. 76, 7949–7956.

Prevention and mitigation strategies for *Campylobacter* with focus on poultry production

6

Thomas Alter

Institute of Food Hygiene, Free University Berlin, Berlin, Germany

6.1 INTRODUCTION

Numerous studies point out that currently a complete elimination of *Campylobacter* in the (poultry) food chain is not feasible in most countries. Truly effective and commonly applicable solutions for the eradication of *Campylobacter* along the food chain are still missing (Wagenaar et al., 2006, 2013). The current aim should be to establish control measures and intervention strategies to reduce and minimize the occurrence of *Campylobacter* spp. in livestock (especially poultry flocks), and to reduce the quantitative *Campylobacter* load in animals and foods. The most efficient intervention measure to control *Campylobacter* spp. in broiler meat is to reduce the *Campylobacter* level on carcasses, rather than reducing the prevalence of *Campylobacter* in broiler flocks (Nauta et al., 2009). Rosenquist et al. (2003) were able to calculate a correlation between the reduction of the quantitative *Campylobacter* contamination of poultry carcasses and a reduction in the incidence of human campylobacteriosis. Simulations showed that a reduction of *Campylobacter* counts on chicken carcasses by two orders of magnitude can lead to a 30-fold decrease in human campylobacteriosis cases. A reduction in the flock prevalence by a factor of 30 would be necessary to achieve the same reduction in the human incidence.

The implementation of efficient regulatory *Campylobacter* control measures in broiler production proved to be useful in New Zealand, and was intensively discussed by a recent EFSA opinion (EFSA, 2011). Recently, Swart et al. (2013) calculated for the Netherlands that the introduction of a process hygiene criterion of, for example, 1000 cfu *Campylobacter* per gram of chicken meat could lead to a reduction of the number of human cases by two-thirds. The costs to the poultry industry to meet such a criterion (estimated at approximately €2 million per year) would be considerably lower than the averted costs of illness (approximately €9 million per year).

To provide a scientific basis for risk management decisions with regard to intestinal *Campylobacter* infections, information from risk assessments, epidemiology and the efficiency of intervention methods have to be taken into account (Mangen et al., 2007). Most of these data are available for poultry and poultry meat. Several publications already investigated the efficiency and the cost-benefit of different intervention methods in the poultry production chain, and included that information into risk assessment models (Havelaar et al., 2007). In addition, information on risk management approaches from Iceland (Stern et al., 2003), Norway (Hofshagen and Kruse, 2005), Denmark, and New Zealand (New Zealand Ministry for Primary Industries, 2013) is available.

In most cases, a combination of different intervention methods from farm-to-fork was proposed or applied. By carrying out cost-effectiveness analysis of different intervention strategies to control *Campylobacter* in the New Zealand poultry supply, Lake et al. (2013) suggest that the most cost-effective interventions are applied in the primary processing stage, including phage-based control in broiler flocks. Even though it is generally agreed that on-farm biosecurity measures need to be enhanced, it is not always clear which specific measure needs to be applied, and the costs can be very high. Obviously, not one method is likely to eliminate *Campylobacter* carriage in broilers on-farm, but a number of methods (single or in combination) can reduce the bacterial load sufficiently so as to have an effect on human health (Wassenaar, 2011).

A number of excellent reviews already described the impact of different mitigation strategies on *Campylobacter* colonization of poultry in primary production, and contamination of carcasses at the harvest level (EFSA, 2011; Klein et al., 2015; Meunier et al., 2016; Newell et al., 2011; Wagenaar et al., 2006).

This chapter tries to summarize current information on control and mitigation strategies along the food chain, including approaches at the postharvest level.

6.2 MITIGATION STRATEGIES AT FARM LEVEL

The most cost-effective interventions are applied in the primary processing stage. In general, preharvest intervention strategies can be divided into three groups (Lin, 2009):

- reduction or elimination of environmental exposure (by biosecurity measures),
- combating *Campylobacter* colonization and minimizing the bacterial load (e.g., by application of bacteriocins or bacteriophages), and
- improving host resistance (vaccines, probiotics, competitive exclusion, stimulating the immune system, genetic selection).

Different studies have performed risk-factor analysis for poultry flock colonization. Increased animal age, number of houses on the farm, production type, stocking density, flock size, number of houses on site, presence of other animals on the farm, partial depopulation (thinning), or type of nipple drinkers are associated with the degree of colonization (EFSA, 2011; Näther et al., 2009).

The common denominator explaining most of these divergent risk factors is biosecurity, or the lack of such measures (Wagenaar et al., 2013). In principle, a high biosecurity level on a poultry farm may prevent the introduction of *Campylobacter* into a poultry flock. However, it does not guarantee a *Campylobacter*-free flock at slaughter.

6.2.1 **BIOSECURITY MEASURES**

Biosecurity measures in primary production are designed to prevent the exposure of noninfected animals with *Campylobacter*, or minimize colonization. A variety of animate and inanimate vectors supporting the entry of *Campylobacter* into broiler flocks has been identified.

Many biosecurity measures can be in conflict with other goals of sustainable farming, for example, outdoor farming, free-range farming, open farms, or other animal species on the farm (Klein et al., 2015).

Hygiene barriers have to be practically implemented, with minimum requirements of boot dips, changing footwear, hand washing facilities, and physical barriers. If consequently applied, hygiene barriers can contribute to reduce the risk of infection up to 50% (Gibbens et al., 2001; Newell et al., 2011). Such hygiene barriers seem especially important when there are other livestock on the farm.

Insects (especially flies) acting as a vector for *Campylobacter* transmission were identified as an important risk factor (EFSA, 2011). Hald et al. (2008) estimated that about 30,000 flies are entering a stable through the ventilation system during one broiler rearing cycle. Thus, the application of fly screens showed promising results (Hald et al., 2007). Nonetheless, it has yet to be demonstrated if these data, generated in the Nordic countries, can be transferred as such to warmer countries, since prevalence of *Campylobacter* and occurrence of flies differ.

Restricting access of personnel and visitors to poultry farms or individual poultry houses can reduce the risk of poultry colonization. EFSA (2011) noted that this risk increased with the increase of the number of people working at the farm, or visiting the farm.

Newell et al. (2011) concluded that, in most instances, biosecurity on conventional broiler farms needs to be enhanced further, and this will contribute to the reduction of flock colonization. However, complementary, nonbiosecurity-based approaches are also required to maximize the reduction of *Campylobacter*-positive flocks at farm level.

6.2.2 **VACCINATION**

Proof of the principle that *Campylobacter* antibodies induced by vaccination in chickens can have protective properties has already been shown, and different vaccination strategies have been tested (de Zoete et al., 2007). But, so far, no effective commercially available vaccines are available. This approach failed mainly due to the antigenic diversity of *Campylobacter* strains colonizing animals. Crucial in developing vaccines is the selection of an adequate antigen used. Most research has been focused on flagellin as protective antigen. In research approaches, after application of several subunit or killed vaccines, antibody formation was detectable. However, this could not protect against colonization. On the other hand, developments in human medicine suggest that—by using new methods—these obstacles can be overcome. According de Zoete et al. (2007), three challenges need to be solved to develop an effective vaccine: new antigens that trigger a crossprotection must be identified; a rapid immune response must be inducible with the vaccine; new adjuvants must be developed to further stimulate the immune response to *Campylobacter* spp.

Current studies focused on using *Salmonella* or oocysts of *Eimeria tenella* as vectors delivering antigens (e.g., CjA, Peb1A, Dsp) (Clark et al., 2012; Layton et al., 2011; Wyszynska et al., 2004). These approaches largely resulted in a reduced cecal load of *Campylobacter*, even though colonization could not be eliminated completely.

6.2.3 PASSIVE IMMUNIZATION

Colonization of broilers naturally occurs in 2- to 3-week old chicks, due to the protective effect of maternal antibodies in chick sera in the first weeks of life.

Cawthraw and Newell (2010) demonstrated that, after experimental infection of chicks from *Campylobacter*-positive parents, the colonization was significantly lower than in chicks from *Campylobacter*-negative parents. That phenomenon was only observed in the first days of life. With increasing age, these differences diminished and, from the third week of age, all animals were equally sensitive. This increasing susceptibility parallels the loss of maternally derived, circulating, anti-*Campylobacter*, IgY antibodies. Passive immunization approaches were tested by Hermans et al. (2014): preventive administration of hyperimmune egg yolk significantly reduced *Campylobacter* counts of seeder animals 3 days after oral inoculation. The long-term effect of passive immunization has not been investigated yet.

6.2.4 BACTERIOPHAGES

In vitro and in vivo studies already demonstrated the potential of lytic bacteriophages to control *Campylobacter* along the food chain, where phages can be applied at different stages of production. Most of that research has been focused on preharvest intervention measures. Different phages and application routes were applied in several in vivo trials to evaluate the potential of phages for a reduction of *Campylobacter* in the intestine of poultry (Carvalho et al., 2010; El-Shibiny et al., 2009b; Hammerl et al., 2014; Loc Carrillo et al., 2005; Wagenaar et al., 2005). In most of these studies, *Campylobacter* counts in cecal content could be reduced by an average of 2 $\log_{10}$ units. In addition, the first field trial with *Campylobacter* phages has recently been carried out by Kittler et al. (2013) in commercial broiler flocks. These authors detected a significant reduction of *Campylobacter* of up to 3.2 $\log_{10}$ cfu/g feces by applying a 7.5 $\log_{10}$ pfu dose of a phage cocktail to 10,000–13,500 broilers via drinking water, 6–7 days before slaughter.

Based on their morphology, genome size, and endonuclease restriction profile, the currently known lytic *Campylobacter* phages are divided into three groups (Javed et al., 2014), of which members of group II and group III have already been used for applications in broilers. In general, group II phages have a broader host range than phages of group III, since they frequently infect both *Campylobacter jejuni* and *Campylobacter coli* strains. On the other hand, some group III phages isolated from poultry showed a strong lytic activity on certain *C. jejuni* strains (Loc Carrillo et al., 2005). As proposed by Hammerl et al. (2014), phage cocktails should contain group II as well as group III phages in order to target a broad range of *C. jejuni* and *C. coli* strains. While group III phages infect higher numbers of *C. jejuni* strains than group II phages, *C. coli* strains were exclusively lysed by phages belonging to group II. However, in most field trials, phage cocktails were no more effective than single phages, in terms of reducing *Campylobacter* loads in the intestine but, in principle, they could target more *Campylobacter* strains than a single target-specific phage by using multiple host cell receptors for binding.

Some studies observed an increased phage resistance level after phage treatment. But that had no significant impact on colonization of resistant strains, due to the loss of virulence, and their poor ability to compete with susceptible strains (Connerton et al., 2011).

6.2.5 PROBIOTIC BACTERIA

A number of studies on the use of probiotic bacteria to control *Campylobacter* in poultry have been published, but results are variable (Morishita et al., 1997; Robyn et al., 2013; Santini et al., 2010). Although the use of probiotics (e.g., *Lactobacillus acidophilus* and *Enterococcus faecium*) showed in individual studies a reduction in the colonization of the intestine with *Campylobacter* spp., however, these results could not be verified by other groups. Chaveerach et al. (2004), after having screened *Enterococcus* and *Escherichia coli* strains, identified one *Lactobacillus* strain that showed bactericidal activity against different *Campylobacter* strains in vitro. Similarly, Santini et al. (2010) identified a *Bifidobacterium* strain with anti-*Campylobacter* activities. In vivo testing showed that the *C. jejuni* concentration in poultry feces was significantly reduced in chickens administered with *B. longum* PCB 133.

When testing probiotic strains present in the multispecies probiotic product PoultryStar sol (*E. faecium, Pediococcus acidilactici, B. animalis, L. salivarius*, and *L. reuteri*), Ghareeb et al. (2012) showed that the cecal colonization by *C. jejuni* was significantly reduced by this probiotic treatment at both 8 and 15 days postchallenge.

6.2.6 BACTERIOCINS

Various research groups applied bacteriocins to reduce *Campylobacter* colonization in poultry. Meunier et al. (2016) recently gave an overview of the performed experiments. Some purified bacteriocins reduced *Campylobacter* colonization in poultry: OR-7 (Stern et al., 2006), L-1077 (Svetoch et al., 2011), SMXD51 (Messaoudi et al., 2012), all from *L. salivarius* strains; E-760 and E 50–52 from *Enterococcus* spp. (Line et al., 2008; Svetoch et al., 2008); SRCAM 602 from *Paenibacillus polymyxa* (Stern et al., 2005). These studies show that purified bacteriocins are, in general, more effective to lower *Campylobacter* intestinal load than the application of probiotic strains. Meunier et al. (2016) speculate that a higher concentration is reached by directly applying purified bacteriocins than by administering probiotics releasing bacteriocins.

6.2.7 CHEMICAL FEED AND WATER ADDITIVES

In vitro, Chaveerach et al. (2002) successfully demonstrated the ability of organic acids (formic, acetic, hydrochloric, and propionic acid) to reduce *Campylobacter* spp.

Byrd et al. (2001) supplemented drinking water with 0.44% lactic acid during a 10-h feed withdrawal assay preslaughter. A significant reduction of crop contamination with *Campylobacter* was observed, compared to a control group (62.3% vs. 85.1% prevalence). The prevalence of *Campylobacter* was also reduced on the corresponding prechill carcass rinses after slaughter by 14.7%. When adding organic acids in combination with medium chain fatty acids (MCFA), ammonium formate, and coconut/palm kernel fatty acid distillate to the drinking water, at a concentration of 0.075% of the blended organic acids (in combination with MCFA), the carriage of *Campylobacter* spp. in naturally colonized broiler flocks and in cecum content of broilers was reduced. However, the final *Campylobacter* contamination of the corresponding carcasses after slaughter was not significantly lowered. Monocaprin as water additive showed a *Campylobacter*-reducing effect on cloacal counts in *Campylobacter* colonized broiler chicken, but cecal counts could not be predictably reduced (Hilmarsson et al., 2006; Metcalf et al., 2011).

Solis de los Santos et al. (2010) demonstrated a beneficial effect of caprylic acid added to feed, leading to a reduction up to 2 $\log_{10}$ cfu, compared to a control group.

The discrepancy between in vitro and in vivo results was highlighted by Hermans et al. (2010), who did not detect any effect of three medium-chain fatty acids on cecal *Campylobacter* loads after adding these fatty acids to chicken feed, despite their bactericidal activities in vitro, probably due to the protective effect of the mucus layer, in contrast to a field-study by Jansen et al. (2014).

Summarized, these studies suggest that preharvest application of organic acids and MCFA to drinking water or feed of broilers can lower the cecal carriage in primary production, but large discrepancies between the results of different studies remain.

6.3 MITIGATION STRATEGIES AT HARVEST LEVEL

Strict implementation of biosecurity measures and GMP/HACCP during slaughter and processing is expected to reduce the level of contamination of carcasses and, subsequently, meat. However, such implementations cannot be quantified, since they depend on many interrelated local factors (EFSA, 2011).

Crates and modules used for transportation of poultry to the abattoir are commonly contaminated with *Campylobacter* (due to poor cleaning and disinfection, usually associated with low concentrations of the used disinfectants), and can pose a risk of contamination of broilers. Slader et al. (2002) detected *Campylobacter* in transport crates arriving at the farm, although the transport crates were previously washed at the abattoir. After transportation in these crates, feathers of the birds were contaminated by *Campylobacter*. In addition, transport stress can increase the shedding rate of *Campylobacter* (Whyte et al., 2001).

Feed withdrawal prior to slaughter (usually applied for 8–12 h) will reduce or prevent fecal contamination before and during slaughter. However, a study by Northcutt et al. (2003), investigating the influence of feed withdrawal and slaughter age, showed significant increases in *Campylobacter* numbers by about 0.6 $\log_{10}$ on the carcasses of 56-day-old broilers, but not on 42- and 49-day-old broilers, after feed withdrawal.

Currently, the slaughter process might contribute to intense crosscontamination taking place at different stages of the slaughter line. That phenomenon (higher prevalence on broiler skin than in ceca) has already been described (Hue et al., 2010). Such abattoir effects are related to technological and hygienic standards during slaughter practices that influence cecal and fecal contamination of the carcasses. Heavily contaminated broiler skin samples originate usually from *Campylobacter*-positive flocks. Such heavily contaminated flocks can act as a source of crosscontamination during the slaughter process. Based on data from the EU *Campylobacter* baseline survey, it was concluded that a *Campylobacter*-colonized broiler batch was about 30 times more likely to have the sampled carcass contaminated with *Campylobacter*, compared to a noncolonized batch, and a higher *Campylobacter* count on carcasses was strongly associated with *Campylobacter* colonization of the batch (EFSA, 2010).

To minimize the entry of *Campylobacter* into the abattoir, possible management options at the abattoir are the introduction of logistic slaughter, or "testing and scheduled slaughter." For logistic slaughter, *Campylobacter*-negative flocks are slaughtered at the beginning of each slaughter day. Some studies showed that the prevalence of *Campylobacter* contaminated carcasses and poultry meat can be quite effectively lowered by logistic slaughter (Potturi-Venkata et al., 2007), but it is generally agreed that this measure is only effective when slaughter batches show only a very low *Campylobacter* prevalence (Johannessen et al., 2007; Rosenquist et al., 2003). However, the quantitative effect of logistic slaughter on *Campylobacter* is quite low, since crosscontamination between flocks occurs on a low quantitative level (Reich et al., 2008). Subsequently, in quantitative risk assessments, logistic slaughter showed only a slight effect on the incidence of human campylobacteriosis (Nauta et al., 2009). Promising results were achieved by introducing "testing and scheduling." In this case, flocks with high loads of *Campylobacter* in the gut are identified prior to slaughter, and subsequently the meat derived from these flocks is excluded from the fresh meat market (Havelaar et al., 2007; Rosenquist et al., 2009). Prerequisite for this approach are quick and reliable testing systems that can be applied on farm, at the end of the fattening period, with test results being available prior to slaughter or processing.

At slaughter and processing, fecal leakage should be minimized, and crosscontamination of carcasses reduced. Critical processing steps are scalding, defeathering, evisceration, and chilling. Table 6.1 illustrates the results of intervention measures at different processing steps. At the scalding step, different measures could be applied to avoid crosscontamination: scalding in successive scalding tanks; scalding tanks on a counterflow principle, with a constant supply of fresh water; installation of a brush system before scalding to remove fecal contamination on skin and feathers.

Table 6.1 Intervention Measures Applied at Different Processing Steps During Slaughter and Processing of Poultry

Technological Step	Samples	Measures Applied		Campylobacter Numbers	References
Defeathering	Carcass rinse	Cloacal plugging	Applied	2.52 $\log_{10}$ cfu/mL	Musgrove et al. (1997)
			Control	3.05 $\log_{10}$ cfu/mL	
Scalding	Carcass	Scalding water temperature	53°C	4.5 $\log_{10}$ cfu/carcass	Lehner et al. (2014)
			53.9°C	1.7 $\log_{10}$ cfu/carcass	
	Neck skin	Scalding water temperature	49°C	3.32 $\log_{10}$ cfu/mL	Wempe et al. (1983)
			53°C	2.92 $\log_{10}$ cfu/mL	
			60°C	2.62 $\log_{10}$ cfu/mL	
	Skin	Scalding water temperature	50°C	<1 $\log_{10}$ reduction	Yang et al. (2001)
			60°C	>1 $\log_{10}$ reduction	
Washing	Carcass rinse	Hot water	75°C, 30 s	0.9 $\log_{10}$ cfu/mL reduction	Purnell et al. (2004)
			80°C, 20 s	1.1 $\log_{10}$ cfu/mL reduction	
	Carcass rinse	Chlorinated water (25 ppm), inside outside washer	Prewash	4.69 $\log_{10}$ cfu/mL	Bashor et al. (2004)
			Postwash I	4.38 $\log_{10}$ cfu/mL	
			Postwash II	4.36 $\log_{10}$ cfu/mL	
			Postwash III	4.24 $\log_{10}$ cfu/mL	
	Carcass rinse	Chlorinated water (40 ppm), inside outside washer	Prewash	1.93 $\log_{10}$ cfu/mL	Berrang and Bailey (2009)
			Postwash	1.27 $\log_{10}$ cfu/mL	
Experimental, after evisceration	Skin	Sonosteam (30–40 kHz, 90–94°C for 1–1.5 s)	Pre-Sonosteam	2.35 $\log_{10}$ cfu/g	Musavian et al. (2014)
			Post-Sonosteam	1.40 $\log_{10}$ cfu/g	
	Skin	Hot steam (80°C) for 20 s followed by crust freezing		2.9 $\log_{10}$ cfu/cm^2 reduction	James et al. (2007)

Modified from Klein et al. (2015)

By applying these measures, entry of fecal material into the scalding tanks can be lowered by up to 90%. Increase of the scalding water temperature is a limited option, since product quality can be impaired by increase of temperature. Sensory changes became apparent already after slight temperature changes (Lehner et al., 2014).

To avoid fecal leakage, cloacal plugging was successfully tested (Buhr et al., 2003; Musgrove et al., 1997). However, the technical implementation is difficult, and no commercial systems are available yet.

Equipment for defeathering and evisceration has to be adjusted to bird size. Poorly adjusted equipment can lead to increased lesions or ruptures of the gastrointestinal tract, and subsequently to higher fecal contaminations of the slaughter carcasses (Rosenquist et al., 2006).

Measures to decontaminate chicken carcasses during the slaughter process include physical and chemical methods. Adding chemicals to the washing steps, such as detergents or chlorine formulations, can lead to higher reduction rates of *Campylobacter* on carcasses or poultry products. A variety of chemical decontamination agents were already tested and evaluated by EFSA regarding their safety, such as chlorine dioxide (CD), acidified sodium chlorite (ASC), trisodium phosphate (TSP), and peroxyacetic acid (PAA) in washing treatments, cetylpiridinium chloride and propylene glycol for poultry carcass dipping, and peroxyacetic acid solutions, acetic acid, or lactic acid. However, the possibility of adaptation of *Campylobacter* to acids has been described in vitro (Murphy et al., 2003).

A variety of physical decontamination methods are available. Short carcass treatment with hot water (75–85°C for 20–30 s) resulted in *C. jejuni*-reduction rates of about 1–2 $\log_{10}$ cfu/cm^2 (Purnell et al., 2004; Whyte et al., 2003). A combination of hot steam for 10 s or hot wash water (80°C) followed by crust-freezing showed reductions of *Campylobacter* numbers of approximately 3 $\log_{10}$ cfu/cm^2 breast skin (James et al., 2007). By combining ultrasound and steam (Sonosteam procedure) treatment, *Campylobacter* contamination on carcasses can be reduced by 2.5 $\log_{10}$ cfu/mL (Hansen and Larsen, 2007; Musavian et al., 2014). Many studies investigated the impact of carcass freezing on the *Campylobacter* load, leading to reduction rates of 1–2.2 $\log_{10}$ levels (El-Shibiny et al., 2009a; Sampers et al., 2010; Solow et al., 2003). Often, "crust-freezing" is applied, where the product is quickly cooled by liquid CO_2 or N_2, and coated with ice crystals. Iceland even introduced a compulsory freezing of carcasses from *Campylobacter*-positive slaughter batches (Stern et al., 2003). In general, freezing of meat lowers the quantitative load of *Campylobacter*, however, it does not lead to a complete elimination. After a 4-month freeze time, *Campylobacter* spp. could still be detected in 80% of the examined carcasses by culture methods (Sandberg et al., 2005). These microorganisms can survive on and in frozen products for several months, and are largely found in drip water after thawing. A study by Birk et al. (2004) showed that *Campylobacter* exhibits a high survival capacity in such drip water, and might present a problem in later processing.

A disadvantage of many physical decontamination processes—compared to chemical methods—are the associated higher costs (Havelaar et al., 2007). Until now, the mentioned chemical and physical decontamination methods were mostly evaluated at model scale equipment, or only for a limited amount of time. Additional testing under field conditions is necessary to assess the *Campylobacter*-reducing effects on a daily basis (Klein et al., 2015).

The final chilling step is intended to inhibit or slow bacterial growth on carcasses, and increase shelf-life of the product. Huezo et al. (2007) could not detect any significant differences between air and immersion chilling, with respect to the reduction of *Campylobacter* prevalence or concentrations. Air and immersion chilling reduced the *Campylobacter* concentrations by about 1.4 and 1.0 $\log_{10}$ cfu/mL, when no chemical decontamination agents were used. In other studies, *Campylobacter* reduction rates of 0.41 $\log_{10}$ and 0.59 $\log_{10}$ cfu/mL were achieved, when comparing immersion chilling and air chilling, respectively (Berrang et al., 2008).

Altogether, a reduction of the *Campylobacter* concentration on poultry carcasses by a factor of up to 2 $\log_{10}$ can be accomplished by optimizing the process steps and/or application of control options in the slaughter line (Lindqvist and Lindblad, 2008; Rosenquist et al., 2003).

6.4 MITIGATION STRATEGIES AT POSTHARVEST LEVEL

So far, intervention measures along the food chain have been concentrated at preharvest and harvest level, but a growing number of mostly experimental data on postharvest intervention measures are available.

Campylobacter appear sensitive to various measures of food conservation, such as acidification, drying, or salting. Higher temperatures, such as those obtained in cooking or frying, can kill *Campylobacter* effectively and quickly. *D* values in ground chicken meat at 49, 53, and 57°C were 20.5, 4.85, and 0.79 min, respectively (Blankenship and Craven, 1982). Studies showed that the death rate of *Campylobacter* spp. is logarithmic in the temperature range of 48.8–55.1°C, with *D* values being much lower than those described for *Salmonella* spp. and *Listeria* spp. However, during thermal stress studies above 56°C, Moore and Madden (2000) observed deviations from logarithmic reduction to nonlogarithmic reduction of cells, where a tailing effect was detected by a deviation from linearity in the log survivor curves, suggesting that a heat-resistant subpopulation within its parent strain might exist.

Storage at low temperatures of approximately 4°C for 3–7 days results in a reduction in *Campylobacter* cell count of 0.34–0.81 $\log_{10}$ cfu/g in ground chicken, and of 0.31–0.63 $\log_{10}$ cfu/g in chicken skin (Bhaduri and Cottrell, 2004). Comparable results were published by Blankenship and Craven (1982): an initial inoculum of 6–7 $\log_{10}$ cfu/g of artificially contaminated chicken meat, stored at 4°C, was reduced by about 1–2 $\log_{10}$ units/g over a 17 days period (compared to 2.5–5 $\log_{10}$ units/g reductions at a storage temperature of 23°C).

Freezing of meat lowers the quantitative load of *Campylobacter*. However, it does not lead to a complete elimination. Freeze–thawing cycles significantly reduce the survival, but *C. jejuni* remained viable for at least three freeze–thaw cycles, regardless if frozen at –70°C or –20°C (Lee et al., 1998). Different factors, like ice crystal formation, ice nucleation and dehydration, have been implicated in the freeze-induced injury of bacterial cells. Oxidative stress has been shown to contribute to the freeze–thawing induced killing of *Campylobacter* (Stead and Park, 2000). The majority of studies investigating the susceptibility of *Campylobacter* to freezing in meat detected decreases of between 1 and 3 $\log_{10}$ units within the first days of storage. *Campylobacter* concentrations will largely remain at the reduced level until the end of the storage period (Moorhead and Dykes, 2002).

Studies by Boysen et al. (2007) and Rajkovic et al. (2010) describe the effect of different gaseous mixtures on the survival of *Campylobacter* on poultry meat. The authors conclude that a gas mixture of 70% O_2 and 30% CO_2 significantly reduced the number of *Campylobacter* on poultry meat by up to 2 $\log_{10}$ cfu, within 6–8 days, under cold storage. When stored under 70% N_2 and 30% CO_2 atmosphere over the period of 10 days, no reduction in *Campylobacter* numbers was detectable. In this context, however, it must be pointed out that, under an atmosphere of high O_2 concentration, the growth of spoilage organisms is supported, leading to a shortened shelf-life. The optimum gaseous mixture for achieving the combined objectives of reducing *Campylobacter* and extending shelf was therefore 40–30–30 CO_2–O_2–N_2, which achieved a shelf-life in excess of 14 days (Meredith et al., 2014).

Survival rates of *C. jejuni* in plain marinade (storage temperature 4°C, pH 4.5, sodium chloride content of 5.9%) were described by Björkroth (2005). The inoculum level of 5.4 $\log_{10}$ cfu/mL was reduced by 2.4 $\log_{10}$ cfu/mL within 24 h. After 48 h no *Campylobacter* were detectable. However, when investigating marinated meat products inoculated with 4–5 $\log_{10}$ cfu/mL, these authors were able to detect *Campylobacter* for 9 days, and linked the survival capacity in marinated meat products to the buffering capability of meat that quickly neutralizes the pH of the acidic marinade.

UV irradiation or the application of pulsed light to decontaminate poultry meat or contact surfaces led to a quantitative reduction of *Campylobacter* by several log steps (Haughton et al., 2012, 2011). UV irradiation (0192 J/cm^2) or the application of the high-intensity light pulse (HILP) technology (3 Hz, max. 505 J/pulse) led to reductions of 0.76 $\log_{10}$ and 1.22 $\log_{10}$ cfu/cm^2, respectively. That phenomenon was highly strain-dependent. Gamma-irradiation doses of 1.0 and 1.8 kGy rendered meat free of *Campylobacter* (Lewis et al., 2002).

An alternative method to produce stable food products is high hydrostatic pressure processing (HPP) subjecting liquid and solid foods, with or without packaging, to 100–800 MPa, resulting in adverse effects on microbial physiology. Solomon and Hoover (2004) applied different pressures at ambient temperatures for 10 min to inoculated chicken purée. *C. jejuni*-changes in cell count fitted well among other gram-negative bacteria, and *C. jejuni* numbers were reduced by 2–3 $\log_{10}$ units at 300–325 MPa. A pressure treatment of 400 MPa inactivated *C. jejuni* in food completely. Martinez-Rodriguez and MacKey (2005a,b) noticed that early stationary phase cells were more resistant to high hydrostatic pressure than those in the exponential phase, an unexpected observation given the lack of similar changes in resistance to heat and oxidative stress. Even though pressure resistance varies considerably between the different *Campylobacter* species, and among strains within a species, the risks of *Campylobacter* surviving commercial pressure treatments are very small.

As mentioned earlier, most of the research on phage application to reduce *Campylobacter* has been focused on preharvest intervention measures. Nonetheless, single studies investigated the postharvest treatment, largely on raw or cooked meat. Some reductions in *C. jejuni* numbers on chicken skin or meat have been observed when treated with a corresponding phage, even though single studies could not detect significant reductions at chilling temperatures (Orquera et al., 2012). Larger reductions were achieved by Atterbury et al. (2003) who combined phage application with freezing, resulting in $<2.5 \log_{10}$ cfu reductions. When phages were applied without freezing, the reduction rate decreased to approximately 1 $\log_{10}$ cfu. Data from Bigwood et al. (2008) showed that experiments performed at 24°C (to simulate room temperature) resulted in higher reduction rates of *C. jejuni* of up to 5 $\log_{10}$ units than experiments at 5°C. Since the minimal growth temperature of *C. jejuni/C. coli* is approximately 30°C, lower temperatures prevent bacterial growth and replication of phages. Based on that fact, Connerton et al. (2011) assumed that the reduction of *Campylobacter* numbers under these circumstances is mainly caused by lysis from without.

The fermentation process in raw fermented poultry meat sausages quickly lead to a strong reduction of *Campylobacter* levels in this product. Only a very high experimental inoculum level (4–6 $\log_{10}$ cfu/g) allowed the survival of a low number of *Campylobacter* during the fermentation and ripening process. Lower initial *Campylobacter* inoculums were completely eliminated within a few hours (Alter et al., 2006). That reduction during ripening and fermentation is largely linked to the combined negative impact of the decreasing pH-values, potential suppression by a competitive flora, oxygen exposure, and the diminishing water activity with increased ripening time, in cooperation with the addition of sodium nitrite and sodium chloride (Uyttendaele et al., 1999).

6.5 CONCLUSIONS

Truly effective and commonly applicable solutions for the eradication of *Campylobacter* along the food chain are still missing. Thus, the present aim should be to establish control measures and mitigation strategies to reduce or minimize the occurrence of *Campylobacter* spp. in livestock (especially poultry flocks) and to reduce the quantitative *Campylobacter* load in animals and foods. To this end, a combination of intervention measures at different stages of the food chain appears most promising. At primary production, strict implementation of on-farm biosecurity measures is needed. These intervention strategies have to be accompanied by additional control measures to reduce transmission along the food chain. By optimizing slaughter and processing steps, the *Campylobacter* concentration on slaughter carcasses (and subsequently meat) can be reduced (e.g., by reduction of fecal contamination during slaughter, postharvest decontamination, such as freezing, crust freezing, or usage of chemical agents). Until now, many physical and chemical decontamination measures were mostly evaluated at laboratory scale or model scale equipment. Additional testing of these measures under field conditions is crucial to assess the impact of the individual measures on *Campylobacter* (Klein et al., 2015).

Last but not least, public health authorities should provide targeted consumer advice, and set up education campaigns to raise the awareness on *Campylobacter* infection and the importance of different sources (e.g., food, environment) in the epidemiology of *Campylobacter.*

REFERENCES

Alter, T., Bori, A.I., Hamedy, A., Ellerbroek, L., Fehlhaber, K., 2006. Influence of inoculation levels and processing parameters on the survival of *Campylobacter jejuni* in German style fermented turkey sausages. Food Microbiol. 23, 701–706.

Atterbury, R.J., Connerton, P.L., Dodd, C.E., Rees, C.E., Connerton, I.F., 2003. Application of host-specific bacteriophages to the surface of chicken skin leads to a reduction in recovery of *Campylobacter jejuni*. Appl. Environ. Microbiol. 69, 6302–6306.

Bashor, M.P., Curtis, P.A., Keener, K.M., Sheldon, B.W., Kathariou, S., Osborne, J.A., 2004. Effects of carcass washers on *Campylobacter* contamination in large broiler processing plants. Poult. Sci. 83, 1232–1239.

Berrang, M.E., Bailey, J.S., 2009. On-line brush and spray washers to lower numbers of *Campylobacter* and *Escherichia coli* and presence of *Salmonella* on broiler carcasses during processing. J. Appl. Poult. Res. 18, 74–78.

Berrang, M.E., Meinersmann, R.J., Smith, D.P., Zhuang, H., 2008. The effect of chilling in cold air or ice water on the microbiological quality of broiler carcasses and the population of *Campylobacter*. Poult. Sci. 87, 992–998.

Bhaduri, S., Cottrell, B., 2004. Survival of cold-stressed *Campylobacter jejuni* on ground chicken and chicken skin during frozen storage. Appl. Environ. Microbiol. 70, 7103–7109.

Bigwood, T., Hudson, J.A., Billington, C., Carey-Smith, G.V., Heinemann, J.A., 2008. Phage inactivation of foodborne pathogens on cooked and raw meat. Food Microbiol. 25, 400–406.

Birk, T., Ingmer, H., Andersen, M.T., Jorgensen, K., Brondsted, L., 2004. Chicken juice, a food-based model system suitable to study survival of *Campylobacter jejuni*. Lett. Appl. Microbiol. 38, 66–71.

Björkroth, J., 2005. Microbiological ecology of marinated meat products. Meat Sci. 70, 477–480.

Blankenship, L.C., Craven, S.E., 1982. *Campylobacter jejuni* survival in chicken meat as a function of temperature. Appl. Environ. Microbiol. 44, 88–92.

Boysen, L., Knochel, S., Rosenquist, H., 2007. Survival of *Campylobacter jejuni* in different gas mixtures. FEMS Microbiol. Lett. 266, 152–157.

Buhr, R.J., Berrang, M.E., Cason, J.A., 2003. Bacterial recovery from breast skin of genetically feathered and featherless broiler carcasses immediately following scalding and picking. Poult. Sci. 82, 1641–1647.

Byrd, J.A., Hargis, B.M., Caldwell, D.J., Bailey, R.H., Herron, K.L., McReynolds, J.L., Brewer, R.L., Anderson, R.C., Bischoff, K.M., Callaway, T.R., Kubena, L.F., 2001. Effect of lactic acid administration in the drinking water during preslaughter feed withdrawal on *Salmonella* and *Campylobacter* contamination of broilers. Poult. Sci. 80, 278–283.

Carvalho, C.M., Gannon, B.W., Halfhide, D.E., Santos, S.B., Hayes, C.M., Roe, J.M., Azeredo, J., 2010. The in vivo efficacy of two administration routes of a phage cocktail to reduce numbers of *Campylobacter coli* and *Campylobacter jejuni* in chickens. BMC Microbiol. 10, 232.

Cawthraw, S.A., Newell, D.G., 2010. Investigation of the presence and protective effects of maternal antibodies against *Campylobacter jejuni* in chickens. Avian Dis. 54, 86–93.

Chaveerach, P., Keuzenkamp, D.A., Urlings, H.A., Lipman, L.J., van Knapen, F., 2002. In vitro study on the effect of organic acids on *Campylobacter jejuni/coli* populations in mixtures of water and feed. Poult. Sci. 81, 621–628.

Chaveerach, P., Lipman, L.J., van Knapen, F., 2004. Antagonistic activities of several bacteria on in vitro growth of 10 strains of *Campylobacter jejuni/coli*. Int. J. Food Microbiol. 90, 43–50.

Clark, J.D., Oakes, R.D., Redhead, K., Crouch, C.F., Francis, M.J., Tomley, F.M., Blake, D.P., 2012. *Eimeria* species parasites as novel vaccine delivery vectors: anti-*Campylobacter jejuni* protective immunity induced by *Eimeria tenella*-delivered CjaA. Vaccine 30, 2683–2688.

Connerton, P.L., Timms, A.R., Connerton, I.F., 2011. *Campylobacter* bacteriophages and bacteriophage therapy. J. Appl. Microbiol. 111, 255–265.

de Zoete, M.R., Van Putten, J.P., Wagenaar, J.A., 2007. Vaccination of chickens against *Campylobacter*. Vaccine 25, 5548–5557.

EFSA, 2010. Scientific opinion on quantification of the risk posed by broiler meat to human campylobacteriosis in the EU. EFSA J. 8, 1437.

EFSA, 2011. Scientific opinion on *Campylobacter* in broiler meat production: control options and performance objectives and/or targets at different stages of the food chain. EFSA J. 9, 2105.

El-Shibiny, A., Connerton, P., Connerton, I., 2009a. Survival at refrigeration and freezing temperatures of *Campylobacter coli* and *Campylobacter jejuni* on chicken skin applied as axenic and mixed inoculums. Int. J. Food Microbiol. 131, 197–202.

El-Shibiny, A., Scott, A., Timms, A., Metawea, Y., Connerton, P., Connerton, I., 2009b. Application of a group II *Campylobacter* bacteriophage to reduce strains of *Campylobacter jejuni* and *Campylobacter coli* colonizing broiler chickens. J. Food Prot. 72, 733–740.

Ghareeb, K., Awad, W.A., Mohnl, M., Porta, R., Biarnes, M., Bohm, J., Schatzmayr, G., 2012. Evaluating the efficacy of an avian-specific probiotic to reduce the colonization of *Campylobacter jejuni* in broiler chickens. Poult. Sci. 91, 1825–1832.

Gibbens, J.C., Pascoe, S.J., Evans, S.J., Davies, R.H., Sayers, A.R., 2001. A trial of biosecurity as a means to control *Campylobacter* infection of broiler chickens. Prev. Vet. Med. 48, 85–99.

Hald, B., Skovgard, H., Pedersen, K., Bunkenborg, H., 2008. Influxed insects as vectors for *Campylobacter jejuni* and *Campylobacter coli* in Danish broiler houses. Poult. Sci. 87, 1428–1434.

Hald, B., Sommer, H.M., Skovgard, H., 2007. Use of fly screens to reduce *Campylobacter* spp. introduction in broiler houses. Emerg. Infect. Dis. 13, 1951–1953.

Hammerl, J.A., Jäckel, C., Alter, T., Janzcyk, P., Stingl, K., Knuver, M.T., Hertwig, S., 2014. Reduction of *Campylobacter jejuni* in broiler chicken by successive application of group II and group III phages. PLoS One 9, e114785.

Hansen, D., Larsen, B.S., 2007. Reduction of *Campylobacter* on chicken carcasses by sonostream treatment. Proceedings of European Congress of Chemical Engineering, ECCE-6, Copenhagen, September 16–20, 2007.

Haughton, P.N., Grau, E.G., Lyng, J., Cronin, D., Fanning, S., Whyte, P., 2012. Susceptibility of *Campylobacter* to high intensity near ultraviolet/visible 395+/−5nm light and its effectiveness for the decontamination of raw chicken and contact surfaces. Int. J. Food Microbiol. 159, 267–273.

Haughton, P.N., Lyng, J.G., Cronin, D.A., Morgan, D.J., Fanning, S., Whyte, P., 2011. Efficacy of UV light treatment for the microbiological decontamination of chicken, associated packaging, and contact surfaces. J. Food Prot. 74, 565–572.

Havelaar, A.H., Mangen, M.J., de Koeijer, A.A., Bogaardt, M.J., Evers, E.G., Jacobs-Reitsma, W.F., van Pelt, W., Wagenaar, J.A., de Wit, G.A., van der, Z.H., Nauta, M.J., 2007. Effectiveness and efficiency of controlling *Campylobacter* on broiler chicken meat. Risk Anal. 27, 831–844.

Hermans, D., Martel, A., van Deun, K., Verlinden, M., van Immerseel, F., Garmyn, A., Messens, W., Heyndrickx, M., Haesebrouck, F., Pasmans, F., 2010. Intestinal mucus protects *Campylobacter jejuni* in the ceca of colonized broiler chickens against the bactericidal effects of medium-chain fatty acids. Poult. Sci. 89, 1144–1155.

Hermans, D., Van Steendam, K., Verbrugghe, E., Verlinden, M., Martel, A., Seliwiorstow, T., Heyndrickx, M., Haesebrouck, F., de Zutter, L., Deforce, D., Pasmans, F., 2014. Passive immunization to reduce *Campylobacter jejuni* colonization and transmission in broiler chickens. Vet. Res. 45, 27.

Hilmarsson, H., Thormar, H., Thrainsson, J.H., Gunnarsson, E., Dadadottir, S., 2006. Effect of glycerol monocaprate (monocaprin) on broiler chickens: an attempt at reducing intestinal *Campylobacter* infection. Poult. Sci. 85, 588–592.

Hofshagen, M., Kruse, H., 2005. Reduction in flock prevalence of *Campylobacter* spp. in broilers in Norway after implementation of an action plan. J. Food Prot. 68, 2220–2223.

Hue, O., Le Bouquin, S., Laisney, M.J., Allain, V., Lalande, F., Petetin, I., Rouxel, S., Quesne, S., Gloaguen, P.Y., Picherot, M., Santolini, J., Salvat, G., Bougeard, S., Chemaly, M., 2010. Prevalence of and risk factors for *Campylobacter* spp. contamination of broiler chicken carcasses at the slaughterhouse. Food Microbiol. 27, 992–999.

Huezo, R., Northcutt, J.K., Smith, D.P., Fletcher, D.L., Ingram, K.D., 2007. Effect of dry air or immersion chilling on recovery of bacteria from broiler carcasses. J. Food Prot. 70, 1829–1834.

James, C., James, S.J., Hannay, N., Purnell, G., Barbedo-Pinto, C., Yaman, H., Araujo, M., Gonzalez, M.L., Calvo, J., Howell, M., Corry, J.E., 2007. Decontamination of poultry carcasses using steam or hot water in combination with rapid cooling, chilling or freezing of carcass surfaces. Int. J. Food Microbiol. 114, 195–203.

Jansen, W., Reich, F., Klein, G., 2014. Large-scale feasibility of organic acids as a permanent preharvest intervention in drinking water of broilers and their effect on foodborne *Campylobacter* spp. before processing. J. Appl. Microbiol. 116, 1676–1687.

Javed, M.A., Ackermann, H.W., Azeredo, J., Carvalho, C.M., Connerton, I., Evoy, S., Hammerl, J.A., Hertwig, S., Lavigne, R., Singh, A., Szymanski, C.M., Timms, A., Kropinski, A.M., 2014. A suggested classification for two groups of *Campylobacter* myoviruses. Arch. Virol. 159, 181–190.

Johannessen, G.S., Johnsen, G., Okland, M., Cudjoe, K.S., Hofshagen, M., 2007. Enumeration of thermotolerant *Campylobacter* spp. from poultry carcasses at the end of the slaughterline. Lett. Appl. Microbiol. 44, 92–97.

Kittler, S., Fischer, S., Abdulmawjood, A., Glünder, G., Klein, G., 2013. Effect of bacteriophage application on *Campylobacter jejuni* loads in commercial broiler flocks. Appl. Environ. Microbiol. 79, 7525–7533.

Klein, G., Jansen, W., Kittler, S., Reich, F., 2015. Mitigation strategies for *Campylobacter* spp. in broiler at pre-harvest and harvest level. Berl. Münch. Tierärztl. Wochenschr. 128, 132–140.

Lake, R.J., Horn, B.J., Dunn, A.H., Parris, R., Green, F.T., McNickle, D.C., 2013. Cost-effectiveness of interventions to control *Campylobacter* in the New Zealand poultry meat food supply. J. Food Prot. 76, 1161–1167.

Layton, S.L., Morgan, M.J., Cole, K., Kwon, Y.M., Donoghue, D.J., Hargis, B.M., Pumford, N.R., 2011. Evaluation of *Salmonella*-vectored *Campylobacter* peptide epitopes for reduction of *Campylobacter jejuni* in broiler chickens. Clin. Vaccine Immunol. 18, 449–454.

Lee, A., Smith, S.C., Coloe, P.J., 1998. Survival and growth of *Campylobacter jejuni* after artificial inoculation onto chicken skin as a function of temperature and packaging conditions. J. Food Prot. 61, 1609–1614.

Lehner, Y., Reich, F., Klein, G., 2014. Influence of process parameter on *Campylobacter* spp. counts on poultry meat in a slaughterhouse environment. Curr. Microbiol. 69, 240–244.

Lewis, S.J., Velasquez, A., Cuppett, S.L., McKee, S.R., 2002. Effect of electron beam irradiation on poultry meat safety and quality. Poult. Sci. 81, 896–903.

Lin, J., 2009. Novel approaches for *Campylobacter* control in poultry. Foodborne Pathog. Dis. 6, 755–765.

Lindqvist, R., Lindblad, M., 2008. Quantitative risk assessment of thermophilic *Campylobacter* spp. and cross-contamination during handling of raw broiler chickens evaluating strategies at the producer level to reduce human campylobacteriosis in Sweden. Int. J. Food Microbiol. 121, 41–52.

Line, J.E., Svetoch, E.A., Eruslanov, B.V., Perelygin, V.V., Mitsevich, E.V., Mitsevich, I.P., Levchuk, V.P., Svetoch, O.E., Seal, B.S., Siragusa, G.R., Stern, N.J., 2008. Isolation and purification of enterocin E-760 with broad antimicrobial activity against gram-positive and gram-negative bacteria. Antimicrob. Agents Chemother. 52, 1094–1100.

Loc Carrillo, C., Atterbury, R.J., el Shibiny, A., Connerton, P.L., Dillon, E., Scott, A., Connerton, I.F., 2005. Bacteriophage therapy to reduce *Campylobacter jejuni* colonization of broiler chickens. Appl. Environ. Microbiol. 71, 6554–6563.

Mangen, M.J., Havelaar, A.H., Poppe, K.P., de Wit, G.A., 2007. Cost-utility analysis to control *Campylobacter* on chicken meat: dealing with data limitations. Risk Anal. 27, 815–830.

Martinez-Rodriguez, A., MacKey, B.M., 2005a. Factors affecting the pressure resistance of some *Campylobacter* species. Lett. Appl. Microbiol. 41, 321–326.

Martinez-Rodriguez, A., MacKey, B.M., 2005b. Physiological changes in *Campylobacter jejuni* on entry into stationary phase. Int. J. Food Microbiol. 101, 1–8.

Meredith, H., Valdramidis, V., Rotabakk, B.T., Sivertsvik, M., McDowell, D., Bolton, D.J., 2014. Effect of different modified atmospheric packaging (MAP) gaseous combinations on *Campylobacter* and the shelf-life of chilled poultry fillets. Food Microbiol. 44, 196–203.

Messaoudi, S., Kergourlay, G., Dalgalarrondo, M., Choiset, Y., Ferchichi, M., Prevost, H., Pilet, M.F., Chobert, J.M., Manai, M., Dousset, X., 2012. Purification and characterization of a new bacteriocin active against *Campylobacter* produced by *Lactobacillus salivarius* SMXD51. Food Microbiol. 32, 129–134.

Metcalf, J.H., Donoghue, A.M., Venkitanarayanan, K., Reyes-Herrera, I., Aguiar, V.F., Blore, P.J., Donoghue, D.J., 2011. Water administration of the medium-chain fatty acid caprylic acid produced variable efficacy against enteric *Campylobacter* colonization in broilers. Poult. Sci. 90, 494–497.

Meunier, M., Guyard-Nicodeme, M., Dory, D., Chemaly, M., 2016. Control strategies against *Campylobacter* at the poultry production level: biosecurity measures, feed additives and vaccination. J. Appl. Microbiol. 120, 1139–1173.

Moore, J.E., Madden, R.H., 2000. The effect of thermal stress on *Campylobacter coli*. J. Appl. Microbiol. 89, 892–899.

Moorhead, S.M., Dykes, G.A., 2002. Survival of *Campylobacter jejuni* on beef trimmings during freezing and frozen storage. Lett. Appl. Microbiol. 34, 72–76.

Morishita, T.Y., Aye, P.P., Harr, B.S., Cobb, C.W., Clifford, J.R., 1997. Evaluation of an avian-specific probiotic to reduce the colonization and shedding of *Campylobacter jejuni* in broilers. Avian Dis. 41, 850–855.

Murphy, C., Carroll, C., Jordan, K.N., 2003. Induction of an adaptive tolerance response in the foodborne pathogen *Campylobacter jejuni*. FEMS Microbiol. Lett. 223, 89–93.

Musavian, H.S., Krebs, N.H., Nonboe, U., Corry, J.E., Purnell, G., 2014. Combined steam and ultrasound treatment of broilers at slaughter: a promising intervention to significantly reduce numbers of naturally occurring campylobacters on carcasses. Int. J. Food Microbiol. 176, 23–28.

Musgrove, M.T., Cason, J.A., Fletcher, D.L., Stern, N.J., Cox, N.A., Bailey, J.S., 1997. Effect of cloacal plugging on microbial recovery from partially processed broilers. Poult. Sci. 76, 530–533.

Näther, G., Alter, T., Martin, A., Ellerbroek, L., 2009. Analysis of risk factors for *Campylobacter* species infection in broiler flocks. Poult. Sci. 88, 1299–1305.

Nauta, M., Hill, A., Rosenquist, H., Brynestad, S., Fetsch, A., van der Logt, P., Fazil, A., Christensen, B., Katsma, E., Borck, B., Havelaar, A., 2009. A comparison of risk assessments on *Campylobacter* in broiler meat. Int. J. Food Microbiol. 129, 107–123.

New Zealand Ministry for Primary Industries, 2013. *Campylobacter* Risk Management Strategy 2013–2014, Wellington.

Newell, D.G., Elvers, K.T., Dopfer, D., Hansson, I., Jones, P., James, S., Gittins, J., Stern, N.J., Davies, R., Connerton, I., Pearson, D., Salvat, G., Allen, V.M., 2011. Biosecurity-based interventions and strategies to reduce *Campylobacter* spp. on poultry farms. Appl. Environ. Microbiol. 77, 8605–8614.

Northcutt, J.K., Berrang, M.E., Dickens, J.A., Fletcher, D.L., Cox, N.A., 2003. Effect of broiler age, feed withdrawal, and transportation on levels of coliforms, *Campylobacter*, *Escherichia coli* and *Salmonella* on carcasses before and after immersion chilling. Poult. Sci. 82, 169–173.

Orquera, S., Golz, G., Hertwig, S., Hammerl, J., Sparborth, D., Joldic, A., Alter, T., 2012. Control of *Campylobacter* spp. and *Yersinia enterocolitica* by virulent bacteriophages. J. Mol. Genet. Med. 6, 273–278.

Potturi-Venkata, L.P., Backert, S., Vieira, S.L., Oyarzabal, O.A., 2007. Evaluation of logistic processing to reduce cross-contamination of commercial broiler carcasses with *Campylobacter* spp. J. Food Prot. 70, 2549–2554.

Purnell, G., Mattick, K., Humphrey, T., 2004. The use of "hot wash" treatments to reduce the number of pathogenic and spoilage bacteria on raw retail poultry. J. Food Eng. 62, 29–36.

Rajkovic, A., Tomic, N., Smigic, N., Uyttendaele, M., Ragaert, P., Devlieghere, F., 2010. Survival of *Campylobacter jejuni* on raw chicken legs packed in high-oxygen or high-carbon dioxide atmosphere after the decontamination with lactic acid/sodium lactate buffer. Int. J. Food Microbiol. 140, 201–206.

Reich, F., Atanassova, V., Haunhorst, E., Klein, G., 2008. The effects of *Campylobacter* numbers in caeca on the contamination of broiler carcasses with *Campylobacter*. Int. J. Food Microbiol. 127, 116–120.

Robyn, J., Rasschaert, G., Hermans, D., Pasmans, F., Heyndrickx, M., 2013. In vivo broiler experiments to assess anti-*Campylobacter jejuni* activity of a live *Enterococcus faecalis* strain. Poult. Sci. 92, 265–271.

Rosenquist, H., Boysen, L., Galliano, C., Nordentoft, S., Ethelberg, S., Borck, B., 2009. Danish strategies to control *Campylobacter* in broilers and broiler meat: facts and effects. Epidemiol. Infect. 137, 1742–1750.

Rosenquist, H., Nielsen, N.L., Sommer, H.M., Norrung, B., Christensen, B.B., 2003. Quantitative risk assessment of human campylobacteriosis associated with thermophilic *Campylobacter* species in chickens. Int. J. Food Microbiol. 83, 87–103.

Rosenquist, H., Sommer, H.M., Nielsen, N.L., Christensen, B.B., 2006. The effect of slaughter operations on the contamination of chicken carcasses with thermotolerant *Campylobacter*. Int. J. Food Microbiol. 108, 226–232.

Sampers, I., Jacxsens, L., Luning, P.A., Marcelis, W.J., Dumoulin, A., Uyttendaele, M., 2010. Performance of food safety management systems in poultry meat preparation processing plants in relation to *Campylobacter* spp. contamination. J. Food Prot. 73, 1447–1457.

Sandberg, M., Hofshagen, M., Ostensvik, O., Skjerve, E., Innocent, G., 2005. Survival of *Campylobacter* on frozen broiler carcasses as a function of time. J. Food Prot. 68, 1600–1605.

Santini, C., Baffoni, L., Gaggia, F., Granata, M., Gasbarri, R., Di Gioia, D., Biavati, B., 2010. Characterization of probiotic strains: an application as feed additives in poultry against *Campylobacter jejuni*. Int. J. Food Microbiol. 141 (Suppl. 1), S98–S108.

Slader, J., Domingue, G., Jorgensen, F., McAlpine, K., Owen, R.J., Bolton, F.J., Humphrey, T.J., 2002. Impact of transport crate reuse and of catching and processing on *Campylobacter* and *Salmonella* contamination of broiler chickens. Appl. Environ. Microbiol. 68, 713–719.

Solis de los Santos, F., Hume, M., Venkitanarayanan, K., Donoghue, A.M., Hanning, I., Slavik, M.F., Aguiar, V.F., Metcalf, J.H., Reyes-Herrera, I., Blore, P.J., Donoghue, D.J., 2010. Caprylic acid reduces enteric *Campylobacter* colonization in market-aged broiler chickens but does not appear to alter cecal microbial populations. J. Food Prot. 73, 251–257.

Solomon, E.B., Hoover, D.G., 2004. Inactivation of *Campylobacter jejuni* by high hydrostatic pressure. Lett. Appl. Microbiol. 38, 505–509.

Solow, B.T., Cloak, O.M., Fratamico, P.M., 2003. Effect of temperature on viability of *Campylobacter jejuni* and *Campylobacter coli* on raw chicken or pork skin. J. Food Prot. 66, 2023–2031.

Stead, D., Park, S.F., 2000. Roles of Fe superoxide dismutase and catalase in resistance of *Campylobacter coli* to freeze–thaw stress. Appl. Environ. Microbiol. 66, 3110–3112.

Stern, N.J., Hiett, K.L., Alfredsson, G.A., Kristinsson, K.G., Reiersen, J., Hardardottir, H., Briem, H., Gunnarsson, E., Georgsson, F., Lowman, R., Berndtson, E., Lammerding, A.M., Paoli, G.M., Musgrove, M.T., 2003. *Campylobacter* spp. in Icelandic poultry operations and human disease. Epidemiol. Infect. 130, 23–32.

Stern, N.J., Svetoch, E.A., Eruslanov, B.V., Kovalev, Y.N., Volodina, L.I., Perelygin, V.V., Mitsevich, E.V., Mitsevich, I.P., Levchuk, V.P., 2005. *Paenibacillus polymyxa* purified bacteriocin to control *Campylobacter jejuni* in chickens. J. Food Prot. 68, 1450–1453.

Stern, N.J., Svetoch, E.A., Eruslanov, B.V., Perelygin, V.V., Mitsevich, E.V., Mitsevich, I.P., Pokhilenko, V.D., Levchuk, V.P., Svetoch, O.E., Seal, B.S., 2006. Isolation of a *Lactobacillus salivarius* strain and purification of its bacteriocin, which is inhibitory to *Campylobacter jejuni* in the chicken gastrointestinal system. Antimicrob. Agents Chemother. 50, 3111–3116.

Svetoch, E.A., Eruslanov, B.V., Levchuk, V.P., Perelygin, V.V., Mitsevich, E.V., Mitsevich, I.P., Stepanshin, J., Dyatlov, I., Seal, B.S., Stern, N.J., 2011. Isolation of *Lactobacillus salivarius* 1077 (NRRL B-50053) and characterization of its bacteriocin, including the antimicrobial activity spectrum. Appl. Environ. Microbiol. 77, 2749–2754.

Svetoch, E.A., Eruslanov, B.V., Perelygin, V.V., Mitsevich, E.V., Mitsevich, I.P., Borzenkov, V.N., Levchuk, V.P., Svetoch, O.E., Kovalev, Y.N., Stepanshin, Y.G., Siragusa, G.R., Seal, B.S., Stern, N.J., 2008. Diverse antimicrobial killing by *Enterococcus faecium* E 50-52 bacteriocin. J. Agric. Food Chem. 56, 1942–1948.

Swart, A.N., Mangen, M.J., Havelaar, A.H., 2013. Microbiological Criteria as a Decision Tool for Controlling *Campylobacter* in the Broiler Meat Chain, RIVM Report 330331008/2013. Dutch Ministry of Health, Welfare and Sport, Bilthoven, The Netherlands.

Uyttendaele, M., De Troy, P., Debevere, J., 1999. Incidence of *Salmonella, Campylobacter jejuni, Campylobacter coli,* and *Listeria monocytogenes* in poultry carcasses and different types of poultry products for sale on the Belgian retail market. J. Food Prot. 62, 735–740.

Wagenaar, J.A., Bergen, M.A., Mueller, M.A., Wassenaar, T.M., Carlton, R.M., 2005. Phage therapy reduces *Campylobacter jejuni* colonization in broilers. Vet. Microbiol. 109, 275–283.

Wagenaar, J.A., French, N.P., Havelaar, A.H., 2013. Preventing *Campylobacter* at the source: why is it so difficult? Clin. Infect. Dis. 57, 1600–1606.

Wagenaar, J.A., Mevius, D.J., Havelaar, A.H., 2006. *Campylobacter* in primary animal production and control strategies to reduce the burden of human campylobacteriosis. Rev. Sci. Tech. 25, 581–594.

Wassenaar, T.M., 2011. Following an imaginary *Campylobacter* population from farm to fork and beyond: a bacterial perspective. Lett. Appl. Microbiol. 53, 253–263.

Wempe, J.M., Genigeorgis, C.A., Farver, T.B., Yusufu, H.I., 1983. Prevalence of *Campylobacter jejuni* in two California chicken processing plants. Appl. Environ. Microbiol. 45, 355–359.

Whyte, P., Collins, J.D., McGill, K., Monahan, C., O'Mahony, H., 2001. The effect of transportation stress on excretion rates of campylobacters in market-age broilers. Poult. Sci. 80, 817–820.

Whyte, P., McGill, K., Collins, J.D., 2003. An assessment of steam pasteurization and hot water immersion treatments for the microbiological decontamination of broiler carcasses. Food Microbiol. 20, 111–117.

Wyszynska, A., Raczko, A., Lis, M., Jagusztyn-Krynicka, E.K., 2004. Oral immunization of chickens with avirulent *Salmonella* vaccine strain carrying *C. jejuni* 72Dz/92 cjaA gene elicits specific humoral immune response associated with protection against challenge with wild-type *Campylobacter*. Vaccine 22, 1379–1389.

Yang, H., Li, Y., Johnson, M.G., 2001. Survival and death of *Salmonella typhimurium* and *Campylobacter jejuni* in processing water and on chicken skin during poultry scalding and chilling. J. Food Prot. 64, 770–776.

Legal aspects and microbiological criteria for *Campylobacter* spp. in the food processing chain

Felix Reich, Günter Klein

Institute of Food Quality and Food Safety, University of Veterinary Medicine Hannover, Foundation, Hannover, Germany

7.1 INTRODUCTION

Campylobacteriosis is a disease of importance for humans and for society, due to the costs resulting from treatment and healthcare, as well as from the loss of human resources (Chapter 2). These conditions resulted in the inclusion of *Campylobacter* and campylobacteriosis in the monitoring and surveillance systems of developed countries. A disease is classified as notifiable when it is considered to be important, for example, because of its frequency of appearance, or because the severity of symptoms demand data collection and regular observation. In such cases, is mandatory by law to report the disease to the authorities. Risk assessment and risk management strategies usually rely on the quantification of its severity, and on costs for society. Source localization and baseline studies may identify sources in case of foodborne diseases, and help set priority targets. Finally, microbiological criteria can eventually be set at various points in the food chain where implementation seems most promising in order to lower consumer exposure, by forcing processors of foods to implement countermeasures, and to prevent contamination.

7.2 NOTIFICATION OF CAMPYLOBACTERIOSIS

Campylobacteriosis in humans is under surveillance in several countries, with an established health surveillance system. There are differences in the kind of data collection on cases, and in the way of reporting. For example, human campylobacteriosis is notifiable in Canada commencing with 1986 (Public Health Agency of Canada, 2015), in Europe (notifiable as causative agent), and in New Zealand/Australia (New Zealand Ministry of Health, 2014; Australian Department of Health, 2015). In the USA, campylobacteriosis was not nationally notifiable until 2015; however, up to that time, states could still decide to report cases to the National Notifiable Diseases Surveillance

Campylobacter. http://dx.doi.org/10.1016/B978-0-12-803623-5.00007-1

System (NNDSS; Center for Disease Control and Prevention, 2015). The notification is in general addressed to the national health protection agency, or the equivalent authorities in a country. The reporting parties are healthcare practitioners, laboratories, clinics, etc. Data includes: the reporting person/institution together with the reporting date, data on the patient, and data on the case. The data in question can be different in various countries, and it can be quite extensive, as seen in one example in Fig. 7.1.

Foodborne Disease Reporting Form

(form can be used to report Amebiasis, *Campylobacter* spp., *Cryptosporidium* spp., *Cyclospora* spp., *E. coli* infection, *Giardia, Listeria* spp., *Salmonella* spp., *Shigella* spp., Trichinosis, *Vibrio* spp., *Yersinia* spp.)

> **Disease Specific Information**

Disease Name: _____ Onset date: _____ Reporting date :___ / ___ / ____

> **Patient Demographic Information**

Last name: _____ First name: _____ Middle name: _____

Date of birth: ___ / ___ / ____ Patient age: _____ Medical record #: _____

Preferred Language: ☐ English ☐ Other: _____

Country of birth: ☐ United States ☐ Other: _____ ☐ Unknown

Gender: ☐ Male ☐ Female ☐ Transgender ☐ Unknown

Address: _____ County: _____

City: _____ State: ____ Zip: _____ ☐ Address unknown ☐ Homeless

Phone: _____ Alternate phone: _____

Occupation: _____ Parent/guardian name: _____

Ethnicity: ☐ Hispanic/Latino ☐ Non-Hispanic/Non-Latino ☐ Unknown

Race (check all that apply): ☐ American Indian/Alaskan Native ☐ Asian ☐ Native Hawaiian/Pacific Islander ☐ White
☐ Black/African American ☐ Unknown ☐ Other: _____

> **Hospital/Clinic Information**

Reporter name: _____ Reporting institution: _____

Ordering provider: _____ Provider phone: _____

Lab: _____ Lab phone: _____

Who should MDH contact if additional information is needed:

☐ Reporter ☐ Provider ☐ Lab ☐ Other: _____

Specimen collection date: _____ Specimen source: _____

Lab result date: _____

Hospitalized: ☐ Yes ☐ No ☐ Unknown If yes, admit date: ___ / ___ / ____

Hospital name: _____ Discharge date: ___ / ___ / ____

Died: ☐ Yes ☐ No If yes, date of death: _____

Pregnant (if applicable): ☐ No ☐ Yes, due date: ___ / ___ / ____

> **Foodborne Disease Specific Information**

Foodhandler: ☐ Yes ☐ No ☐ Unknown If yes, restaurant name: _____

Childcare attendee/worker: ☐ Yes ☐ No ☐ Unknown If yes, childcare center name: _____

Antibiotics prescribed: ☐ Yes ☐ No ☐ Unknown If yes, antibiotic name: _____

Antibiotic treatment date: ___ / ___ / ____

Did the patient travel outside the United States one week prior to illness onset: ☐ Yes ☐ No ☐ Unknown

Did the patient develop hemolytic uremic syndrome (HUS): ☐ Yes ☐ No ☐ Unknown (If yes, please complete HUS form)

MDH Minnesota Dept. of Health
625 N Robert St.
St. Paul, MN 55164
Phone: 651-201-5414 I Fax: 651-201-9007

5/2015

FIGURE 7.1 Example of a Reporting Form for Notifiable Diseases by the Minnesota Department of Health (Minnesota Department of Health, 2015)

In contrast, the presence of *Campylobacter* in animals or livestock is not in general notifiable, with the exception of bovine genital campylobacteriosis (*C. fetus* ssp. *venerealis*), listed by OIE (OIE, 2015) as a disease of importance to international trade. However, campylobacteriosis in other animals, when caused by thermophilic *Campylobacter* spp., is reportable in Germany (Federal Ministry of Justice and Consumer Protection, 1983), for example.

Besides notification of cases and corresponding epidemiological data, selected isolates from monitoring programs are tested for antimicrobial resistance. In Europe, member states have to collect isolates of *C. jejuni* from broilers and turkey at slaughter, in order to test for antimicrobial resistance when yearly production exceeds 10,000 t of slaughter weight (European Commission, 2013).

7.3 MICROBIOLOGICAL CRITERIA

Several sources of human infection with *Campylobacter* have been identified over the past decade. Most of them are food related (MPI, 2014; EFSA, 2005; Greig and Ravel, 2009). As a result, a proposed way of decreasing the burden of human *Campylobacter* infections is to reduce consumer exposure by reducing numbers of *Campylobacter* in food. This can be achieved by different targets along the food processing chain, applying various different strategies from farm to food prepared at home (Chapter 6). The aim of the measures taken is to improve food safety, and to increase protecting consumer health. For this task, it is important to identify the risk that a food harbors, which, in the case of ready-to-eat foods, is dependent on the amount of pathogenic bacteria present.

Microbiological criteria are one metric of microbiological risk management (MRM), as outlined by Codex Alimentarius (CAC, 2007). The general principles of establishing a microbiological criterion as presented in "Principles and guidelines for the establishment and application of microbiological criteria related to foods" by Codex Alimentarius (CAC, 2013a) are as follows:

- ensuring an appropriate level of consumer health protection and considering fair trade practices;
- practicability and feasibility, establishing only when necessary;
- clearly articulating the purpose of the criterion;
- criterion should be based on scientific information and analysis;
- inclusion of knowledge of the microorganism in regards to occurrence and behavior in the food chain;
- considering intended and actual use of the final product by the consumer;
- periodic review of the criterion to evaluate it being still relevant.

The purpose of a microbiological criterion can be to decide the acceptance or rejection of a food lot. It can also include: verifying (1) the performance of the food safety control system, (2) the microbiological status of the food in relation to criteria of acceptance, (3) control measures meeting performance/food safety objectives. It can also inform food business operators on microbiological levels when applying best practices (CAC, 2013a). As a risk management metric, microbiological criteria

Table 7.1 Definitions of Risk Assessment Metrics in Context With Microbiological Criteria (CAC, 2003, 2013b)

ALOP	Appropriate level of protection is the level of protection deemed appropriate by the country establishing a sanitary measure to protect human life or health within its territory. (This concept may otherwise be referred to as the "acceptable level of risk")
FSO	Food safety objective: the maximum frequency of a hazard in a food at the time of consumption that provides or contributes to the appropriate level of protection (ALOP)
PC	Performance criterion: the effect in frequency and/or concentration of a hazard in a food at the time of consumption that provides or contributes to the appropriate level of protection (ALOP)
PO	Performance objective: the maximum frequency and/or concentration of a hazard in a food at a specified step in the food chain before the time of consumption that provides or contributes to an FSO or ALOP, as applicable

have to be viewed at in context with other MRM related metrics, such as the appropriate level of protection (ALOP), food safety objective (FSO), performance criterion (PC), and performance objective (PO), see Table 7.1 for definitions.

The components and other considerations regarding a microbiological criterion are as follows. Firstly, the purpose is clearly outlined as above. Secondly, it concerns the food, or process it applies to. In this case the criterion is meant to be applied to *Campylobacter* in poultry. The point of application needs to be specified; for *Campylobacter* in poultry it is usually at the end of the slaughtering, for example, after chilling (FSIS, 2015a; EFSA, 2011). Following that, a limit has to be defined. Furthermore, a sampling plan is needed, defining the number (*n*) of samples to collect under the criterion, and defining whether an acceptance number (*c*) is needed, taking also into account the size of the sample. Finally, the analytical methods for testing need to be defined. In the case of *Campylobacter*, the current method is ISO 10272 for detection or enumeration (ISO, 2006a, 2006b), but there can be other similar methods, as proposed by authorities as mandatory.

7.4 MICROBIOLOGICAL CRITERIA FOR *CAMPYLOBACTER* IN EUROPE, NEW ZEALAND, AND IN THE USA

In Europe, microbiological criteria for food were set in 2005 (EC, 2005), where safety criteria are defined for *Salmonella* or *Listeria monocytogenes* in foods that are put on the market. Process hygiene criteria such as the aerobic colony count or the number of *Escherichia coli*, for example, on meat after processing, were established in order to evaluate the hazard analysis and critical control point (HACCP)-based processing and general hygiene concepts.

Campylobacter was identified by the European Food Safety Authority (EFSA) as a microbiological hazard, specifically in poultry, that is not recognizable by the

Table 7.2 Proposed Harmonized Epidemiological Indicators for *Campylobacter* in Poultry Processing (EFSA, 2012b)

Indicators (Animal/Food/ Category/Other)	Food Chain Stage	Analytical/ Diagnostic Method	Specimen
HEI[a] 1 *Campylobacter* in poultry flocks prior to slaughter (2–3 days prior slaughter)	Farm	Microbiology— real-time PCR	Cecal droppings
HEI[a] 2 Controlled housing conditions at farm for poultry flocks (including biosecurity)	Farm	Auditing	Not applicable
HEI[a] 3 Use of partial depopulation in the flock	Farm	Food chain information	Not applicable
HEI[a] 4 *Campylobacter*	Slaughterhouse	Microbiology— enumeration	Cecal content
HEI[a] 5 *Campylobacter* in birds-carcass after slaughter process and chilling	Slaughterhouse	Microbiology— enumeration	Neck and breast skin

[a]HEI, harmonized epidemiological indicator.

official control of the meat at slaughterhouses (EFSA, 2012a). Following these assessments, harmonized epidemiological indicators (HEI) were proposed, defined as "prevalence or concentration of the hazard at a certain stage of the food chain or an indirect indicator of the hazards (such as audits of farms or evaluation of process hygiene) that correlates to human health risk caused by the hazard" (EFSA, 2012b). These should be implemented in broiler meat processing (Table 7.2). Harmonized epidemiological indicator number 5, testing for *Campylobacter* at the end of broiler slaughtering and chilling, is intended as a data collection tool for subsequently setting a quantitative *Campylobacter* target at slaughterhouse level. Bearing in mind such targets, a *Campylobacter* criterion can be implemented for broilers: for example, < 1000 cfu/g of neck skin, also taking into account possible intervention strategies at various stages of poultry meat processing (EFSA, 2011, 2012b). A potential microbiological criterion could be applicable for individual food batches, when placed on the market. For broiler meat, a baseline survey allowed a possible public health risk reduction of > 50% at European Union level, if all batches of fresh meat complied with the criterion of 1000 cfu/g of neck or breast skin. When a critical limit of 500 cfu/g of neck and breast skin was evaluated, even a health risk reduction of < 90 % could be expected (EFSA, 2011). It was calculated that the limit for a health risk reduction of > 50% would lead to 15% of all batches in the European Union to not comply with the criteria, while the > 90% target would result in 45% noncompliance. Different critical limits (m), different number of samples analyzed (n), and the definition of the number of samples allowed not complying per batch (c) would have a different impact on the potential of risk reduction, but also on the

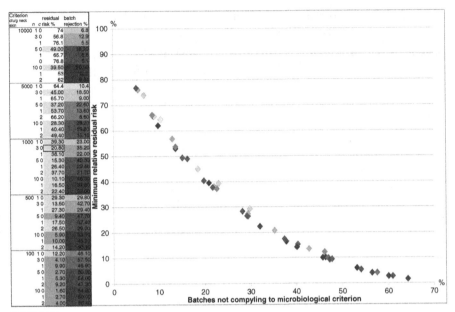

FIGURE 7.2

Estimations for batch rejection and residual risk in percentage, based on various *Campylobacter* criteria in neck skin (100, 500, 1,000, 5,000, or 10,000 cfu/g of neck skin; *n* = 1, 3, 5, 10; *c* = 0, 1, 2). Data is based on the European weighted mean (EFSA, 2011). Colors in the table correspond to the colored diamonds in the figure.

rate of noncomplying batches (Fig. 7.2). The analytical method for the enumeration of *Campylobacter* in neck skin or neck and breast skin is ISO 10272 (ISO, 2006b).

Besides the discussions at the European Food Safety Authority (EFSA) regarding a microbiological criterion for *Campylobacter* at European level, a national *Campylobacter* reduction target was already agreed on in the United Kingdom (FSA, 2015). The rationale was that according to the baseline survey on *Campylobacter* in broiler carcasses in the European Union in 2008 (EFSA, 2010), the prevalence in UK broiler carcasses was above the European average. The target was planned to be quantitative, set upon a three-class plan, by classifying *Campylobacter* categories from < 100 cfu/g of neck skin, 100–1000 cfu/g, and > 1000 cfu/g – aiming for the reduction of the amount of highly contaminated broilers (> 1000 cfu/g neck skin) after the chilling process to below 10% by 2015 (Table 7.3).

The Netherlands discussed the implementation of a *Campylobacter* process hygiene criterion (PHC). The critical limits set have to be evaluated under the perspective of risk management aspects. Of course, a stricter criterion would result in a higher consumer risk reduction, but also in a higher ratio of noncomplying broiler batches. The cost for the industry of about €2 million is in contrast to the reduced costs-of-illness of €9 million, and the resulting public health benefit of about

Table 7.3 *Campylobacter* Target in the United Kingdom (FSA, 2015)

	Campylobacter Count Categories		
	< 100 cfu/g	100–1000 cfu/g	> 1,000 cfu/g
Baseline 2008	42%	31%	27%
Target 2015	Expected improvement	Expected improvement	10% target

Baseline 2008: UK percentage distribution of broilers in the three Campylobacter count categories, according to the European baseline study of 2008 (EFSA, 2010); Target 2015: reduction of the percentage of broilers corresponding to the > 1000 cfu/g Campylobacter count category to at least 10% by 2015. For the two other categories, improvement to the baseline rate was expected.

400 healthy life-years (Swart et al., 2013). Indeed, the evaluation of this model shows a cost benefit for society, but the generation of costs for the industry. Therefore, it is necessary to see how the cost could be distributed, for example, through price adjustments, or institutional measures such as taxes that would allow for equal competition between countries (Swart et al., 2013).

Microbiological criteria in Europe are defined in commission regulation (EC) 2073/2005 (Commission Regulation, 2005), and can be classified into food safety criteria (FSC) and process hygiene criteria (PHC). Both have different functions. An FSC defines the acceptability of a food or batch of foodstuffs placed on the market. Not meeting an FSC would usually result into withdrawal from the market, or recall of a food when already placed on the market. An alternative would be submitting the food to further processing in order to eliminate the hazard. A PHC assists in indicating the acceptable functioning of the production process when the set values are met. It is therefore not applicable to foods placed on the market, but defines a threshold after which corrective actions are required to improve processing hygiene (Commission Regulation, 2005; HPA, 2009).

In 2006, there were 15,873 cases of human campylobacteriosis notified in New Zealand (> 350 cases/100,000 people) (ESR, 2011). The New Zealand *Campylobacter* risk management strategy was then implemented in 2007, aimed at reducing human foodborne campylobacteriosis cases by 50%, over a 5-year period, from 2008 to 2012 (ESR, 2011). Poultry meat was identified to be the primary exposure pathway for foodborne campylobacteriosis, so a mandatory performance target was established at the end of primary processing of chicken meat in 2008 (MPI, 2013) During the course of this period, campylobacteriosis cases decreased to below 8000 cases/year (< 200 cases/100,000 people per year). This level was then to be maintained throughout 2012–2014. Efforts were made to identify the contribution of other sources and pathways important for human infections (MPI, 2013). The legal requirements for *Campylobacter* testing in poultry are national specifications, according to a *Campylobacter* performance target (CPT) (MPI, 2012). This demands daily sampling in commercial slaughterhouses (> 1 million broilers produced/year) and weekly sampling in small slaughterhouses (< 1 million broilers produced/year). Carcass rinse samples were chosen and evaluated under a moving window approach

Table 7.4 *Campylobacter*-Performance Target (CPT), New Zealand (MPI, 2012)

	Comercial Slaughterhouses[a]	Small Slaughterhouses[b]
Moving window of three processing periods Noncompliance	45 samples during 15 processing days[c] EF $n \geq 7$; DF ≥ 30	9 samples over 3 processing weeks[d] EF ≥ 2 ; DF ≥ 6

EF: Enumeration failure when n samples > 3.78 log_{10} cfu/carcass; DF: detection failure when n samples > 2.3 log_{10} cfu/carcass.
[a]> 1 million broiler/year.
[b]< 1 million broiler/year.
[c]1 processing period covers 5 processing days with 3 samples per day ≥ 15 samples.
[d]1 processing period covers one week with 3 samples per week ≥ 3 samples.

(Table 7.4). The moving window covers three consecutive processing periods. The target has two components: a detection limit (2.3 log_{10} cfu/carcass) and an enumeration limit (3.78 log_{10} cfu/carcass). In case of either or both an enumeration failure (EF) and/or a detection failure (DF), the window is counted as noncompliant. Decisions are made upon the number of consecutive noncompliant windows. The noncompliance database is reset to zero when compliance to both EF and DF during a window is reestablished. Measures after consecutive noncompliance start with initiating corrective actions, followed by external audits, and can include sanctions such as prolonged official supervision, and even closure of the premises is possible after at least eight consecutive noncompliant windows (MPI, 2015). The analytic method is applied to whole carcasses analyzed by rinsing after the chilling process. The method is based on the ISO 10272 direct plating method, and no enrichment is applied to the carcass rinse samples (MPI, 2012).

In the USA, performance standards were initially set for the reduction of *Salmonella* in poultry based on verification programs that started in 1996, under the "Pathogen reduction; hazard analysis and critical control point (HACCP) systems" final rule, published as FSIS regulation 310.25 (FSIS, 1996). After revision in 2010, these standards now also include *Campylobacter* legally (FSIS, 2010a). These were revised in 2012 to also include, besides carcasses at slaughter establishments, ground or comminuted not ready to eat meat from chicken and turkey, and to define the sampling regime and analytical methods (FSIS, 2015a). The newly proposed pathogen reduction performance standards of 2015 are part of "The Healthy People 2020" (ODPHP, 2015) goal that aims to reduce by 33% human illness attributed to *Campylobacter*, by 2020. The standards are implemented as a weekly sampling over 52 weeks, and evaluation of compliance as a moving window approach started on May 1, 2015 (Table 7.5). Samples of whole carcass rinses, comminuted meat, or meat parts are to be collected according to FSIS directive 10,250.1 (FSIS, 2013) complemented by FSIS notices number 16-15 (FSIS, 2015b), and number 31-15 (FSIS, 2015c). Samples of poultry carcasses, of comminuted poultry meat, and of chicken parts are to be analyzed according to USDA method MLG 41.03 (FSIS, 2014). Finally, establishments will be categorized in three categories (I–III), depending on the compliance

Table 7.5 Performance Standards for *Campylobacter* in Poultry Selected by FSIS and Estimated Impact on Illness Reduction in the USA After Implementation of Product Standards (FSIS, 2015a)

Compliance Fraction[a] (%)	Metric	Broiler Carcasses	Turkey Carcasses	Raw Chicken Parts	Not Ready-to-Eat Comminuted	
					Chicken	Turkey
	Performance standards	8 of 51	3 of 56	4 of 52	1 of 52	1 of 52
50	Reduction of illness	—	—	32%	37%	19%
40	Reduction of illness	—	—	27%	30 %	15%
30	Reduction of illness	—	—	20%	22%	11%

[a]*Percentage of reduction of establishments that initially failed the performance standards.*

to performance standards (FSIS, 2015a). Category I is defined as reaching of 50% or less of the performance standard, showing "consistent process control;" category II establishments have more than 50% of the performance standard, but meeting the standard during the 52-week moving window, showing "variable process control;" category III establishments exceed the performance standard, showing "highly variable process control." Noncompliance will result in actions to be taken during meat processing. A compliance guideline was created by FSIS (FSIS, 2010b) to help establishments improve processing conditions and help lowering the amount of *Campylobacter* on the meat when noncompliance was identified.

7.5 EXPECTATIONS AND RESULTS OF PERFORMANCE TARGETS/CRITERIA

The aim of microbiological criteria or performance targets for *Campylobacter* is that by reducing the presence or numbers of *Campylobacter* in poultry meat, there would be a direct reduction of human campylobacteriosis cases. Experience in New Zealand showed an initial drop of human campylobacteriosis after the implementation of the *Campylobacter* strategy in August 2006, and finally the CPT, in April 2008. A reduction of 50% in the number of reported human campylobacteriosis cases was subsequently observed. The main focus of the target was to improve the hygiene in slaughter establishments by reducing the amount of poultry meat with high *Campylobacter* counts (EF), but still taking *Campylobacter* presence into account (DF). Although the percentage of broilers with counts above detection and enumeration limits were reduced during the course of these past years, no additional effect on the number of human campylobacteriosis cases could be observed (MPI, 2015).

The New Zealand Ministry of Primary Industries considered several options: (1) to maintain the status quo for limits, (2) tighten the enumeration and/or detection limits, (3) take additional measures for poor performers, (4) implement additional measures for startup premises (MPI, 2015). Discussions between the industry and officials indicate the preference for maintaining the status quo of limits, and a focus on poor performing establishments to improve their rate of noncompliance to a better overall performance (MPI, 2015).

In Europe, the criterion is planned to be quantitative. Different sampling plans have been proposed, with different calculated benefits to the increase in consumer health. One important point to consider is the difference between European countries that will result in different efforts to meet recommended microbiological limits, and also different rates of noncompliance (EFSA, 2011).

7.6 NGO, RETAIL INITIATIVES, AND READY-TO-EAT FOOD

There exist several guidelines for food trade and marketing for ready-to-eat food. As such food is intended to be eaten directly, without further heat treatment or other handling that would kill pathogenic bacteria, microbiological criteria demand absence (nondetection) of *Campylobacter* in 25 g of sample size (HPA, 2009; Centre for Food Safety, 2007; Food Standards Australia New Zealand, 2001).

REFERENCES

Australian Department of Health, 2015. Australian National Notifiable Diseases and Case Definitions. Australian Department of Health, Canberra. Available from: http://www.health.gov.au/casedefinitions#c

CAC, 2003. Guidelines for Food Import Control Systems. Codex Alimentarius Commission, Codex Alimentarius Guidelines 47.

CAC, 2007. Principles and Guidelines for the Conduct of Microbiological Risk Management (MRM). Codex Alimentarius Commission, Codex Alimentarius Guidelines 63.

CAC, 2013a. Principles and Guidelines for the Establishment and Application of Microbiological Criteria Related to Foods—Revised and Renamed 2013. Codex Alimentarius Commission, Codex Alimentarius Guidelines 21.

CAC, 2013b. Procedural Manual, twenty-first ed. FAO/WHO, Geneva, Switzerland, J.F.W.F.S. Programme.

Center for Disease Control and Prevention, 2015. National Notifiable Conditions. Centers for Disease Control and Prevention, Atlanta, GA. Available from: http://wwwn.cdc.gov/nndss/conditions/campylobacteriosis/

Centre for Food Safety, Hong Kong, 2007. Microbiological Guidelines for Ready-to-Eat Food. Centre for Food Safety, Hong Kong.

The European Commission, 2005. Commission Regulation (EC) No 2073/2005 of 15 November 2005 on Microbiological Criteria for Foodstuffs 2005. The European Commission, Luxembourg, Amtsblatt Nr. L 338 vom 22/12/2005 S. 0001–0026. Available from: http://eur-lex.europa.eu/legal-content/EN/ALL/?uri=CELEX:32005R2073

The European Commission, 2005. Commission Regulation (EC) No 2073/2005 of 15 November 2005 on Microbiological Criteria for Foodstuffs. The European Commission, Official Journal of the European Union, Luxembourg.

EFSA, 2005. Scientific report of the scientific panel on Biological Hazards on the request from the Commission related to *Campylobacter* in animals and foodstuffs annex. EFSA J. 173, 1–105.

EFSA, 2010. Analysis of the baseline survey on the prevalence of *Campylobacter* in broiler batches and of *Campylobacter* and *Salmonella* on broiler carcasses in the EU, 2008, Part A: *Campylobacter* and *Salmonella* prevalence estimates. EFSA J. 8 (3), 100.

EFSA, 2011. Scientific opinion on *Campylobacter* in broiler meat production: control options and performance obectives and/or targets at different stages of the food chain. EFSA J. 9 (4), 141.

EFSA, 2012a. Scientific opinion on the public health hazards to be covered by inspection of meat (poultry). EFSA J. 10 (62741), 179.

EFSA, 2012b. Technical specifications on harmonised epidemiological indicators for biological hazards to be covered by meat inspection of poultry. EFSA J. 10 (62764), 87.

ESR, 2011. Annual Report Concerning Foodborne Disease in New Zealand, 2010. Institute of Environmental Science & Research Limited, Wellington, New Zealand.

The European Commission, 2013. 2013/652/EU: Commission Implementing Decision of 12 November 2013 on the Monitoring and Reporting of Antimicrobial Resistance in Zoonotic and Commensal Bacteria. The European Commission, Luxembourg.

Federal Ministry of Justice and Consumer Protection, 1983. Verordnung über Meldepflichtige Tierkrankheiten. Federal Ministry of Justice and Consumer Protection, Bonn, Germany.

Food Standards Australia New Zealand, 2001. Guidelines for Microbiological Examination of Ready-to-eat Foods. Food Standards Australia New Zealand, Wellington, New Zealand and Kingston. Available from: http://www.foodstandards.gov.au/publications/pages/guidelinesformicrobi1306.aspx

FSA, 2010. The Joint Government and Industry Target to Reduce *Campylobacter* in UK Produced Chickens by 2015. Food Standards Authority, London, UK. Available from: https://www.food.gov.uk/science/microbiology/campylobacterevidenceprogramme

FSIS, 1996. Pathogen Reduction; Hazard Analysis and Critical Control Point (HACCP) Systems. Food Safety and Inspection Service, Department of Agriculture, USDA, Atlanta, GA, Final Rule.

FSIS, 2010a. New Performance Standards for Salmonella and *Campylobacter* in Young Chicken and Turkey Slaughter Establishments; New Compliance Guides. Food Safety and Inspection Service, Department of Agriculture, USDA, Atlanta, GA.

FSIS, 2010b. Compliance Guideline for Controlling Salmonella and *Campylobacter* in Poultry, third ed. FSIS, Atlanta, GA.

FSIS, 2013. *Salmonella* and *Campylobacter* verification program for raw meat and poultry products in 10,250.1. 2013. USDA, Food Safety and Inspection Service, Atlanta, GA.

FSIS, 2014. Isolation and Identification of *Campylobacter jejuni/coli/lari* from Poultry Rinse, Sponge and Raw Product Samples. Laboratory Guidebook 2014. USDA, Food Safety and Inspection Service, Atlanta, GA.

FSIS, 2015a. Changes to the *Salmonella* and *Campylobacter* Verification Testing Program: Proposed Performance Standards for *Salmonella* and *Campylobacter* in Not-Ready-to-Eat Comminuted Chicken and Turkey Products and Raw Chicken Parts and Related Agency Verification Procedures and Other Changes to Agency Sampling. Food Safety and Inspection Service, Department of Agriculture, USDA, Atlanta, GA.

FSIS, 2015b. Raw Chicken Parts Sampling Project, in 16-15. 2015. USDA, Food Safety and Inspection Service, Atlanta, GA.

FSIS, 2015c. Not Ready-to-Eat Comminuted Poultry Sampling Project, in 31-15. USDA, Food Safety and Inspection Service, Atlanta, GA.

Greig, J.D., Ravel, A., 2009. Analysis of foodborne outbreak data reported internationally for source attribution. Int. J. Food Microbiol. 130 (2), 77–87.

HPA, 2009. Guidelines for Assessing the Microbiological Safety of Ready-to-Eat Foods. Health Protection Agency, London, UK.

ISO, 2006a. ISO 10272-1:2006 Microbiology of Food and Animal Feeding Stuffs—Horizontal Method for Detection and Enumeration of *Campylobacter* spp.—Part 1 Detection Method. ISO, Geneva, Switzerland.

ISO, 2006b. ISO/TS 10272-2:2006 Microbiology of Food and Animal Feeding Stuffs—Horizontal Method for Detection and enumeration of *Campylobacter* spp.—Part 2 Colony-Count Technique. ISO, Geneva, Switzerland.

Minnesota Department of Health, 2015. Foodborne Disease Reporting Form. Minnesota Department of Health, Minneapolis, MN. Available from: http://www.health.state.mn.us/divs/idepc/dtopics/reportable/forms/foodborneform.html

MPI, 2012. Animal Products (National Microbiological Database Specifications) Notice 2012. Ministry for Primary Industries, Wellington, New Zealand.

MPI, 2013. *Campylobacter* Risk Management Strategy 2013–2014. Ministry for Primary Industries, Wellington, New Zealand.

MPI, 2014. Source Attribution Studies for Campylobacteriosis in New Zealand. Technical Paper No: 2014/11. Ministry of Primary Industries, Wellington, New Zealand.

MPI, 2015. Review of the Poultry NMD Programme's *Campylobacter* Performance Target (CPT) Limit(s). Ministry of Primary Industries, Wellington, New Zealand.

New Zealand Ministry of Health, 2014. Notifiable Diseases. New Zealand Ministry of Health, Wellington, New Zealand. Available from: http://www.health.govt.nz/our-work/diseases-and-conditions/notifiable-diseases

Office of Disease Prevention and Health Promotion, 2015. Healthy People 2020. ODPHP, Washington, DC. Available from: http://www.healthypeople.gov/2020/topics-objectives/topic/food-safety/objectives

OIE, 2015. OIE-Listed Diseases, Infections and Infestations in Force in 2015. World Organization for Animal Health, Paris, France. Available from: http://www.oie.int/animal-health-in-the-world/oie-listed-diseases-2015/

Public Health of Canada, 2015. List of Nationally Notifiable Diseases. Public Health Agency of Canada, Ottawa, ON. Available from: http://dsol-smed.phac-aspc.gc.ca/dsol-smed/ndis/list-eng.php

Swart, A.N., Mangen, M.J., Havelaar, A., 2013. Microbiological Criteria as a Decision Tool for Controlling *Campylobacter* in the Broiler Meat Chain. Dutch Ministry of Health, Welfare and Sport, Bilthoven, The Netherlands, RIVM Letter Report, 2013.

Index

Printed in the United States
By Bookmasters